M 1 093 53133 5

D1376042

Information Mar

n and

n

The Government Centre for
Information Systems

LONDON: HMSO

Acknowledgements

The assistance of Colin Hookham of Cambridge Computer Consultants and Dan Rickman of Dan Rickman Associates in the preparation of this volume under contract to CCTA is gratefully acknowledged.

For further information regarding CCTA products please contact:

CCTA Library
Rosebery Court
St Andrews Business Park
Norwich
NR7 0HS
01603 704930

Foreword

This volume is part of CCTA's Information Management Library. The volumes in this Library address the effective production, co-ordination, storage, retrieval, dissemination and management of information from internal and external sources in order to improve the performance of an organization. Information is a valuable resource and its use overall should be effectively managed.

Information Management is not concerned simply with Information Technology. Nor is it the exclusive business of the traditional experts, librarians, records managers, information scientists, database administrators, traders and so on. Information Management is both a policy matter for senior managers and a practical management task for information service professionals and practitioners who have responsibility for its implementation.

The Information Management Library provides guidance on the management of organizations' business-related information. The Library coverage includes Information and Data Management, Geographic Information Systems and Electronic Data Interchange.

This volume is in the Geographic Information Systems set of the Information Management Library.

CCTA welcomes customer views on the Information Management Library publications. Please send your comments to:

Customer Services
Information Management Library
CCTA
Rosebery Court
St Andrews Business Park
NORWICH
NR7 0HS

Contents

Annex

1 Introduction

1.1 Background

Many people find the subject of standards difficult to understand, yet standards play a prominent part in almost all aspects of our everyday lives – from electricity sockets to food additives. The world would be a far more complicated and less predictable place without standards. Just a brief glance at things around you will confirm the scope of influence of standards, and the fact that they bring a sense of order, efficiency and economy to a potentially chaotic environment.

This volume is about a specific need for standards – information standards that apply to digital geographic information (GI) and system standards that apply to geographic information systems (GIS). It is also about the emerging GI/GIS market, the conflict between short-term needs and long-term interests, and the open standards opportunity which could underpin development of the GI/GIS industry and help it to achieve its full potential for those who supply or use related systems and services.

The generally accepted definition of a Geographic Information System is that referred to in *Handling Geographic Information*, the report of the committee of enquiry chaired by Lord Chorley:

> *A Geographic Information System is a system for capturing, storing, checking, integrating, manipulating, analysing and displaying data which are spatially referenced to the earth. This is normally considered to involve a spatially referenced computer database and appropriate applications software.*

This volume has been developed by CCTA as a sponsor member of the Association for Geographic Information (AGI). The AGI Standards Steering Committee provides the membership of BSI IST/36 – the British Standards Institution (BSI) Technical Committee on Geographic Information. The AGI is currently undertaking a programme of GI/GIS standard-related work at a national, European and international level, and such GI/GIS-related standards are particularly important to CCTA and its clients for two reasons. First, because there are important economic, technical and legal reasons for public sector organizations to apply standards, and secondly, because the public sector contains some of the largest groups of users of GI and is therefore potentially a significant beneficiary of GIS technology.

This volume provides an explanation of the drive towards GI/GIS standards and encourages readers to participate actively in whatever way they consider appropriate. The worlds of GIS and standards are equally afflicted with technical jargon, and this may in part explain why often GI/GIS standards messages have not always met with universal support and commitment. This volume tries to make GI/GIS standards accessible to readers, whether users, suppliers, standards producers or others. We hope that this volume will serve to break down many of the communication barriers that isolate advocates of GI/GIS standards and help readers to recognize what GI/GIS standards can do for them.

1.2	**Purpose**

The purpose of this volume is to:

- introduce non-technical readers to the subject of IS standards

- provide readers with an understanding of the impact of IS standards

- position GIS in the context of other standards

- describe the current status of UK, European and international standards relating to GI and GIS

- provide guidance on what users should be specifying

- highlight areas in which further standards are needed

- examine factors that affect the success of standards

- express a view of where GI and GIS standards are leading

- encourage suppliers and users to become involved in standards-making.

1.3	**Who should read this volume**	This volume is intended for all staff involved in the management, planning and procurement of geographic information systems or interested in standards issues.

The primary audience for this publication is:

- IS providers who are interested in offering, or may be required to offer, GIS services

- IS strategic planners

- members of organizations or groups representing existing and potential GIS users.

The publication will also be of interest to:

- IS/IT managers in organizations that are interested in the use of geographic information and GIS technology

- the wider GIS community

- the wider IS/IT standards community.

1.4	**Assumed knowledge**	This volume has been written for readers who are unfamiliar with IS standards and standards-making organizations. Whilst readers will benefit from being familiar with geographic information systems and their uses, such knowledge is not considered essential to appreciate the key points and general principles involved.

Readers who are unfamiliar with GI and GIS may find it useful to read CCTA's Information Management Library volume: *An Introduction to Geographic Information Systems*. A general overview of IS standards can be found in CCTA's IS Planning Subject Guides volume: *Standards and IS Strategy*.

1.5	**Structure of volume**	Chapter 2 provides an introduction to the subject of IS standards and places GI/GIS standards within their international context.

Chapter 3 discusses the business impact of GI/GIS standards and sets out the case for further progress.

Chapter 4 describes the current status of GI/GIS-related standards, and also provides guidance on the use of existing GI/GIS standards for public sector procurements.

Chapter 5 provides a vision of the future of GI/GIS markets, identifies priority areas for standards activity, and looks at possible ways forward for GI/GIS standards.

Annex A provides a summary of the BSI IST/36 – AGI standards agenda.

Annex B provides a summary of the CEN TC/287 standards agenda.

Annex C provides a comparison of AGI/BSI and CEN GI/GIS standards activities.

There is a Bibliography and a Glossary of terms used in this publication.

2   Introducing IS standards

2.1   The role of IS standards

The world of information systems (IS) has come a long way over the last fifteen years. Many developments have occurred:

- the cost of IS/IT solutions has fallen

- the power and flexibility of systems has increased

- the variety of software has expanded

- software has become easier to use

- integrated solutions have become a reality

- IS has moved into many new areas of business activities

- the size of IS markets has grown by an order of magnitude

- the tools/methodologies are now much more refined to enable users to define their requirements and 'own' the solutions

- data/information sharing is a much better understood requirement of systems.

In short, the IS industry has progressed from a closeted world of punched cards and air conditioned computer rooms to the convenience of hand held Personal Digital Assistants which (almost) read handwriting and take electronic communications to the person and away from the desk. Many commentators refer to these developments as an information revolution – a description that is by no means misplaced considering the extent of the changes and the pace at which they have occurred. Progress of this sort has been welcomed by those involved in both the use and the supply of IS, and it is a good example of market forces at work.

The rapid development of the IS industry has not, however, been without problems for those who supply and use it. The very same market forces that have supported innovation, competition and lower prices have produced an industry burdened with complex dependencies between application software, hardware and operating environments. This phenomenon is one of the key factors currently inhibiting the growth of general IS markets and is a particular problem for the GIS community.

The emergence of IS standards has played a vital role in reducing some of the destructive side-effects of such growth (such as technical compatibility problems) and in sustaining progress. But IS standards are not just about technical compatibility; they are also about quality, scalability, portability and building lasting value into the investments of systems developers and users. IS standards have therefore become an important consideration for those considering IS procurement, particularly for those developing an integrated IS infrastructure.

2.1.1 Public sector policy

It is UK public sector policy that appropriate IS standards will be used in the planning, development, procurement, operation and the management of information systems. This policy has been adopted to promote business effectiveness and efficiency by using standards to:

- achieve value for money through competition

- maximize choice of solutions

- avoid the risks of being locked into proprietary products and services

- safeguard investments in infrastructure projects

- enable exchange of information between organizations

- minimize the time, effort and cost of developing and managing IS

- improve the quality of information systems and services.

The policy also serves to ensure that the many parts of government comply with national and European legislation and international guidelines.

Many GIS commentators recognize that IS standards should be extended to accommodate the special needs of GIS so that future GIS projects are less constrained by issues such as technical compatibility, uncertain quality, limited scalability, and so on, and they argue that it is only with such standards that markets for both geographic information and GIS will achieve their full potential. The remainder of this publication discusses the nature of the GI/GIS standards initiative, and describes the various issues that affect its future.

2.2   Types of IS standards

Throughout this publication reference is made to different types of standards, each of which is explained below:

- corporate standards

- de facto standards

- de jure standards.

2.2.1   Corporate standards

Any organization – including government departments individually and collectively – can define internal standards for application in its own activities. These may include methods and procedures and are commonly referred to as local, organizational, internal, ad hoc or corporate standards.

Strictly speaking, any type of standard, for example, an international standard, becomes a corporate standard if the organization decides to adopt it as a standard for its own use. However, it is usually more convenient to use the term only for those standards that have no higher status; that is, those intermediate solutions adopted by individuals or organizations in the absence of de facto or de jure standards (see definitions below). In the context of GIS, these would usually be the proprietary standards such as are used for information exchange. For example, Ordnance Survey Transfer Format (OSTF) was used in the early days of digital mapping by a number of organizations because nothing better existed.

Whilst proprietary standards serve a useful purpose in that they ease local compatibility problems, they carry a danger of leading to proprietary lock-in and the prospect of users investing in systems which can become obsolete as the market develops.

Over time proprietary standards may or may not become prominent in the market, in which case they become de facto standards.

2.2.2 De facto standards

De facto standards are standards that are established by the market – often by products which dominate a particular aspect of the market, or which are the first of their kind. Since de facto standards confirm the presence of an established user-base, they provide an incentive for other suppliers to build compatible products, and thereby deliver greater choice to users. One such example is DXF which has become widely adopted for the exchange of graphical data due to the success of the AutoCAD product.

De facto standards can arise for a variety of reasons, and it is important to appreciate that they do not always work in favour of freedom of customer choice. De facto standards can be subject to undue influence by a small number of suppliers (particularly the market leaders), especially where all the details of the standard are not made available and there is no open access.

It is also important to appreciate that de facto standards do not necessarily represent best value for money or best available technology; de facto standards may become established for entirely different reasons. A commonly quoted example from the video industry concerns the VHS video tape format, which was regarded as technically inferior to the Betamax format, but which eventually won through by providing a better selection of video material.

In spite of these potential weaknesses, the presence of a de facto standard is an indication of market confidence and investment in a technology – one of the best indications that the technology, its suppliers and developers will be around for some time; but not necessarily a guarantee of future choice, suitability or economy.

2.2.3 De jure standards

De jure standards by contrast are standards that have been established by wider consultation between users and industry through a formal standards-setting body. In many respects their role is similar to that of de facto standards in that they have the effect of establishing widespread compatibility, but they attempt to achieve this by careful design rather than by exploiting a commercial position, and they also attempt to address other issues such as quality and scalability.

Within the IS industry de jure standards have largely been used to treat the destructive symptoms of rapid market growth and technical diversity. Often the term 'open IS standards' is used to indicate those de jure standards which support the philosophy of markets that are open to free competition or technology which has an ability to communicate and expand openly.

De jure standards operate at national, European and international levels. In the UK the British Standards Institution (BSI) is responsible for the establishment of UK national standards for a wide variety of industries including IS. Other countries have their own national standards organizations, for example, AFNOR, DIN and ANSI. Comité Européen de Normalisation (CEN) is responsible for standards matters on a pan-European basis, and International Organization for Standardization (ISO) at an international level.

Consultation between industry and users can mean that de jure standards have a much greater chance of widespread acceptability and suitability than a proprietary de facto standard. Users generally have more confidence in the technical suitability of de jure standards and there need be less concern about proprietary lock-in.

For IS suppliers de jure standards represent an opportunity to produce products that are known to have widespread acceptability within the user community, are compatible with other products in the market (in this respect), and have no bias towards a particular vendor's solution.

| 2.2.4 | Acceptance timescales | De facto standards become established more quickly than de jure since they are driven by the market which itself moves quickly. Because de facto standards reflect market demand and are based on implemented systems they often make attractive prospects for formal adoption as de jure standards. The Structured Query Language (SQL) standard, widely used in accessing data held in relational databases, is a good example of this. It should be noted, however, that there may |

be differences between the de jure standards specification and what is actually implemented in practice. For example, not all aspects of SQL2 have found acceptance from the vendors, and there is no guarantee at present that SQL2 will be fully implemented as specified in the de jure standard.

Although basing a de jure standard on existing practice has many advantages, de jure standards still take several years to become adopted due to the wide consultation processes and the formal procedures of many standards bodies. It is therefore sometimes necessary for standards bodies to anticipate market demand to ensure that standards are available when required, and some standards-making organizations have developed procedures for fast-tracking standards that are already known to have wide support.

2.2.5   Summary

In summary, de jure standards have an important role to play in delivering technically compatible and suitable business solutions and providing protection from proprietary lock-in. Consequently they are very important to major IS users, such as large corporations and government bodies which face a constant battle to maintain adequate levels of compatibility between IS components – particularly for those organizations needing to establish IS infrastructures which are inherently flexible and able to accommodate diverse forms of IS.

National and European legislation is in place to ensure open competition for major public sector IS procurements, and de jure standards are an important aspect of this legislation.

Unfortunately, there is no guarantee that all de jure standards will become widely accepted. There are de jure standards that have never been widely accepted by the market place, for example, GKS.

The UK government/EU policy on standards is discussed in section 2.4, and expressed in other publications available from CCTA, particularly: *Information Systems Standards – Policy for UK Public Sector Organizations* and *Guide to the requirements of the IT Standards Decision and the Revised Supplies Directive.*

2.3   Hierarchy of standards

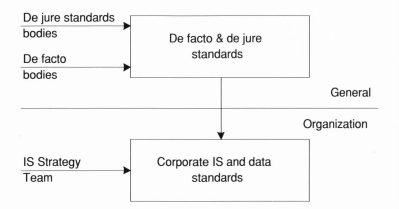

Figure 1: Standards hierarchy

The diagram in Figure 1 indicates a hierarchy of available standards. The aim of standardization is to reduce the effort involved in developing project and organization specific standards.

De facto standards will generally emerge as leaders from the mix of varying corporate and suppliers standards once the market has matured beyond a critical point. It is not yet clear that the GIS market has matured beyond this point, and there is an opportunity for de jure standards to be developed that will anticipate the needs of the future market. This will occur, however, only if such standards:

- are relatively straightforward to implement, that is, easy to use and easy to understand

- have a universally accepted process for verifying that the standards are met; for example, a parser for geographic data in a given format

- gain commitment and acceptance from all parts of the GIS community

- have some expectation of stability. The failure of NTF to gain wider acceptance can be put down in part to its frequent changes from 1.0 to 1.1 then to 2.0. This was probably indicative of the fact that it was developed whilst the market was still too immature.

It is clear that these conditions have not yet been widely fulfilled to date. In particular no data standard in the GI area has provided a verification process.

2.4	Relationship between national, regional and international standards

The organizational relationship between formal standards bodies is generally well established. Due to the constitution of the European Union, all CEN standards are mandatory for European Union member standards institutes. ISO standards are not mandatory, but it is European Union policy to adopt international standards as European standards if they already exist.

It is too early to know what the relationship will be between the ISO and CEN committees on GI/GIS standards. The situation regarding the de jure standards bodies will, however, be affected by the market situation and which, if any, of the de jure standards are implemented and accepted widely by the GIS market.

Standards priorities

Where appropriate European standards exist, public sector organizations should apply these. In the absence of European standards, priority should be given to international standards over national ones. With the prevailing absence of ISO standards in this area, until CEN standards are agreed, the BSI standards are recommended for UK government procurement.

There are currently five cases where reference to standards in procurement is not required:

- where operational continuity of existing systems is needed

- for genuinely innovative projects

- where the standards themselves are now inadequate, because the technology has evolved since the standard was ratified

- where the use of standards would provide an uneconomic solution

- for contracts less than 100,000 ecu in value.

Detailed guidance on the application of IS Standards is available in three CCTA publications:

- *Information Systems Standards: Policy for UK Public Sector Organizations*

- *Guide to the Requirements of the IT Standards Decision and the Revised Supplies Directive*

- the IS Planning Subject Guides volume: *Standards and IS Strategy.*

2.5   Defining the scope of IS standards

Once the need for a new standard is established the scope of standardization must be carefully planned. Unlike de facto IS standards, which often apply to whole products, de jure IS standards only apply to generic aspects of information systems such as for example interfaces and file formats. By concentrating on these aspects, standards can facilitate compatibility without interfering with opportunities to innovate.

Particular care is needed to ensure that programmes of standards work concentrate on advancing a technical field rather than imposing unwanted constraints on it. Consequently, standards makers need to have a very broad understanding of technical issues at a scoping stage, whereas within the development stages they will be required to focus on more detailed issues.

2.5.1   The MUSIC model

CCTA has developed a simple model, known as the MUSIC model, to illustrate the scope of IS standards. The name 'MUSIC model' was derived from the names of topics that open IS standards cover:

- **m**anagement standards

- **u**ser interface standards

- **s**ystems standards

- **i**nformation standards

- **c**ommunications standards.

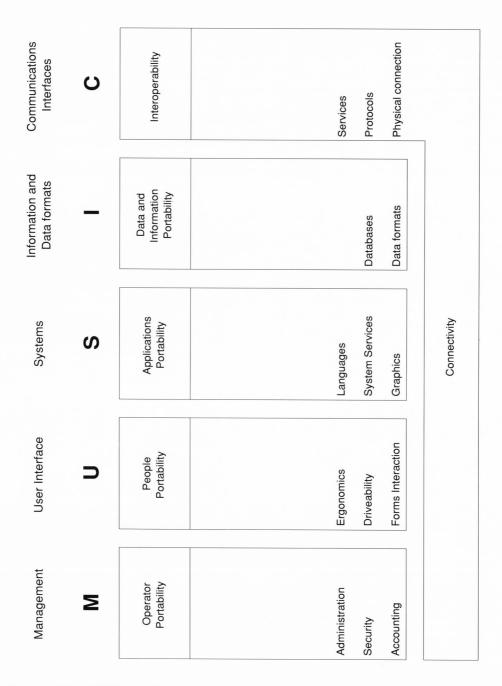

Figure 2: The MUSIC model

The MUSIC model provides a very useful framework within which to consider IS standards (see Figure 2), and has been successfully applied by standards organizations including IEEE, NIST, OSF, X/Open and the EU.

Management standards

Management standards relate to all those functions concerned with procuring, developing, running and monitoring information systems.

These standards focus on the quality, accountability and control of information systems. They do not include the many technical standards which, by bringing rationalization and widespread compatibility, make a major contribution to the management of IS.

A standard relevant to GI(S) in this area is the de facto Open Software Foundation (OSF) Distributed Management Environment (DME) which defines an integrated environment for the user based on managed objects.

User interface standards

User interface standards are concerned with the interface between users and systems; in other words, what you see and what you touch.

Standards relevant to GI(S) in this area are the de facto X Windows, the vendor independent OSF Motif, and the proprietary AT&T/Sun Open Look.

Systems standards

Systems standards embrace most of the standards concerned with the technical characteristics of computer hardware and software.

The majority of systems standards are concerned with technical compatibility. They enable flexibility to respond to changing business needs and they facilitate integration. They also enable an open market, in which choice and competition can act to keep prices down and quality high, and protect investment by diminishing the likelihood that existing systems must be scrapped when new business needs are to be satisfied. System standards help reduce the chances of users being unable to function as a result of hardware and software being no longer supported by the original suppliers and incompatible with products from alternative sources.

Some systems standards are concerned with hardware safety, interference and other aspects largely concerned with protecting third parties. In most cases, compliance with these standards is deemed to satisfy relevant legislation (such as health and safety or electromagnetic interference).

Standards relevant to GI(S) in this area include graphics and operating systems standards, particularly:

- GKS (ISO 7942 – 1985) is the Graphics Kernel Standard – the original standard for display and output devices, devised for programmers to access standard two-dimensional graphics functions.

- PHIGS (ISO 9592) is the Programmer Hierarchical Interactive Graphics System standard, which provides a three-dimensional graphical database accessible by different applications.

- PEX (de facto) is the PHIGS extension to X, which is supported by X Windows and therefore is particularly important, for example, to provide three-dimensional visualization tools for representation of data. It is a de facto standard and may or may not survive long.

- Open GL (de facto) is a three-dimensional graphics language developed by Silicon Graphics.

- The Posix family of operating system standards were originally produced by a group of IEEE committees in the USA. These committees are addressing various aspects, including issues such as standard system functions, administration, security and process communication. The US IEEE 1003.1 standard, which specifies the operating system interface, has been adopted as international standard ISO 9945-1 and this Posix standard has been widely implemented. The rest of the 'Posix family' of IEEE 1003.n standards are in different stages of development and agreement.

Information
standards

Information standards are concerned with the definition, encoding and structuring of information – character sets, data, text, diagrams, images, voice and sound.

Information is a key resource and essential to the operation of the business. Without the observance of common standards, especially for definitions, it is often impossible to move information between applications and to collate information from different sources. The structural changes in business, for example moves towards a federal organization or outsourcing, can seriously endanger the integrity of information as a key resource unless accompanied by the application of standards through executive or consensus action.

Standards relevant to GI(S) in this area are:

- BS 7567 National Transfer Format

- BS 7666 Part 1 Street names, Part 2 Land and Property and Part 3 Addresses

- CORBA

- OLE2

- SQL

- EDI

- EDIFACT.

Communication standards

Communication standards are concerned with the means of communication between systems, over local or wide area networks.

The aim of communication standards is to facilitate both the simple movement of information and the more complex goal of interworking – the co-operation between two systems in order to perform a single task, such as where one is manipulating data held on another system.

Standards relevant to GI(S) in this area are OSI, Posix, TCP/IP.

2.5.2 Profiles

The MUSIC model is a generic one into which a wide range of national, European, and international IS standards will fit. However, many standards cover only one aspect of the model (for example, NTF for Information and Data Formats) and IS users need to be able to establish which standards are appropriate to their requirements.

Work carried out to establish the Open Systems Interconnection (OSI) standard developed the concept of profiles. These profiles represent subsets of standards which collectively define a complete communications environment for an application, for example, the GOSIP (Government OSI Profile).

As with OSI, it has been possible to define a profile of open standards which should apply to GIS-based applications. Figure 3 shows that the MUSIC model can be used to illustrate the context of GIS within existing IS standards, and to define the scope for new GI/GIS standards. The profile is based on the X/Open Common Applications Environment and the EDI-based Open Geomarket developed by the Nordic countries.

This profile has been used to establish the scope for GI/GIS standards work within the BSI IST/36 standards programme referred to in Annex A.

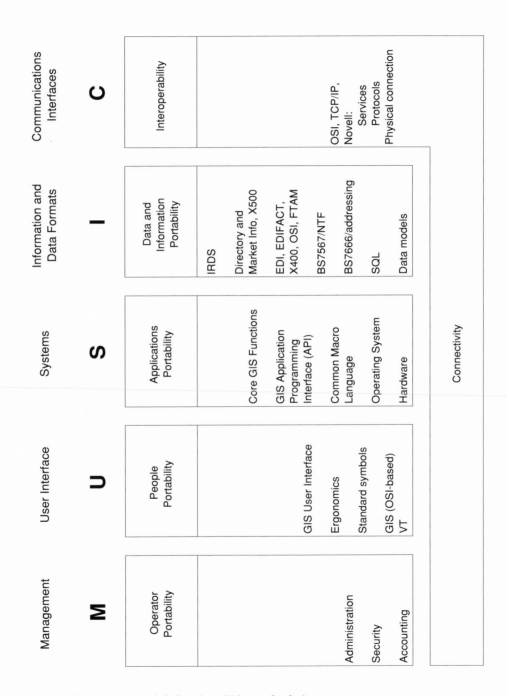

Figure 3: The MUSIC model showing GIS standards in context

3 The role of standards in GI/GIS

Modern GIS products exploit a broad range of new technologies, and organizations will face suitability and compatibility issues to some degree when implementing GIS. It is important to recognize that many, if not most, of the these issues are symptoms of the IS industry generally.

The extension of open standards in general areas of information technology will, in time, alleviate many of the suitability and compatibility problems associated with GIS and help broaden the take-up of GIS. However, there are several areas in which existing IS standards need to be enhanced or GI/GIS specific standards should be developed.

The following sections explain the issues that GI/GIS standards need to address if GIS is to achieve its full potential.

3.1 Data-related issues

One of the special characteristics of geographic information is that it almost always has the potential to be compared and integrated with other geographic information in a meaningful way – using geography as a common theme. This unusual quality is one reason why GIS promises to enhance our understanding greatly of natural and man-made aspects of our environment. However, in practice differing specifications for data formats, quality, spatial units and data models undermine this ability, and significantly hinder useful analysis.

Unless common GI standards are established at organizational, industrial, national, European, or international levels, GIS-based applications have the capacity to misinform and mislead. The GI industry needs to take great care to ensure that confidence in information presented in cartographic form is not damaged by inappropriate arrangements for acquisition, transfer, processing, management and presentation. It therefore needs to build widely acceptable standards on a foundation of experience from geographers, cartographers, statisticians, computer scientists and others, so that GIS achieves broad recognition as a professional business tool.

3.1.1 Data compatibility and portability

A substantial proportion of geographic information is currently available only in the proprietary format of the host/source product. This causes a wide variety of compatibility and portability problems. Often system suppliers and users have to develop bespoke interfaces to translate data from external sources. This can add considerable cost to datasets and there is no guarantee that the full meaning of the data can be reproduced in a target system. The characteristic dependence of GIS on third-party data and the international nature of some applications means that data portability is a widespread problem.

The adoption of a neutral transfer format, such as NTF, limits the number of different formats between which data has to be transferred. With a bilateral approach in contrast the number of translators required is ½ N! (that is, ½ x N x N-1 x N-2 x ... x 1). In other words, there is a factorial increase in the number of translators required as the number of formats increases.

The principal data portability requirement is to establish a recognized language by which data can be exchanged. However, developing a standard for the transfer of GI does not only mean agreeing a syntax by which geographical features can be exchanged, although this is clearly important. It also involves adopting a language which allows the data to be completely reconstructed on another system – including graphical representation, topology and attribute data without any loss of information (and thus value) in the transfer process.

The GIS community currently suffers from a shortage of useful GI and, although GI standards will not in themselves solve this problem, they will serve to encourage availability by making data supply and data purchase easier and a little cheaper.

The adoption of BS 7567 (NTF v2) will help resolve this problem within the UK over the coming years, and further work within CEN and ISO is expected to deliver European and international standards in due course. However, there is evidence that some users will not wait for data transfer standards.

3.1.2 Data quality

Data quality is another area in which new GI standards could play a useful role. Whilst GIS applications can be used in a decision support role, they can be sensitive to the quality of data used, and it can be difficult for users to detect the sorts of errors and inaccuracies that can occur. For example, inaccuracies in the position of county boundaries may not be visibly obvious, but could lead to incorrect analysis of census data.

Future standards for expressing the quality of digital geographic information are likely to involve a number of parameters including:

- completeness

- spatial accuracy

- lineage – including currency, source materials transformations and updates

- attribute accuracy

- topology and connectivity

- logical consistency – fidelity of the data representation in relation to the specified data structures.

3.1.3 Spatial units/
locational references

By its nature GI uses locational references to express the relative, absolute or approximate position of features/objects – latitude/longitude and addresses are two familiar examples, but there are a number of alternatives.

Most GI uses a co-ordinate system of one sort or other. Within the UK most organizations use the National Grid, but other co-ordinate systems are used in Northern Ireland, the Channel Islands and off-shore. New Global Positioning System (GPS)-based surveying tools will inevitably make more geographic data available using the WGS84 co-ordinate system. However, co-ordinates produced from GPS will not match co-ordinates taken from Ordnance Survey maps and there is no simple continuous transformation.

Whilst the use of several co-ordinate systems is inconvenient, data can generally be translated from one to another to allow integration and comparison. The issue becomes much more acute for geographic data represented by irregular spatial units such as counties, regions or postcode areas, and can therefore be a particular problem for handling demographic information.

The government Interdepartmental Group on Geographic Information (IGGI) is currently undertaking work in this area within central government.

3.1.4 Data structures/ models

Another factor that inhibits potential portability, integration and comparison of geographic information is its data model. This problem manifests itself at two levels; first at a product level and secondly at an application level.

Product level

Whilst there is some degree of agreement on the generic elements which should be used to represent geographic information (for example such as lines, polygons, or point features), different GIS products sometimes use slightly different sets of these. For example, not all GIS products support the concept of a grid, and some products use the clockwise/anticlockwise direction of polygon co-ordinates to indicate the difference between islands and holes, that is, the inside and outside of polygon features.

Application level

The way in which data is represented for one application is not necessarily appropriate for another. For example roads may be represented as networks, lines and arcs, or even polygons depending on whether the intended application is route planning, cartography or road maintenance planning, and complex features such the weather need to consider three-dimensional space, the dimension of time, and approximately 50 separate parameters from air pressure to wind speed.

LGMB sponsored definitions

The Local Government Management Board (LGMB) has sponsored and managed the definition of a range of standards for referencing streets, land and property, and addresses. Further projects for standards relating to rights of way and highway locational referencing are ongoing.

The modelling of geographic information for specific applications is an area in which much greater research is needed.

**3.2 System-related
     issues**

These issues encompass compatibility with existing IS, and
the portability, usability and interworking of application
systems.

3.2.1 Compatibility with
       existing IS
       infrastructure

For those involved in planning GIS for large organizations,
achieving and maintaining an appropriate level of
compatibility with existing IS can be a very real and
expensive problem. Existing IT hardware, operating systems,
Human-Computer Interface (HCI), networks, databases and
peripherals each constrain free choice for new IS purchases.
Open standards policies are therefore a prominent feature of
IS strategies with major public and private sector
organizations.

Although some GIS products have been designed with an
inherent ability to adapt to different operating environments,
the prevailing situation is that user choice of GIS products
remains severely restricted by technical compatibility and
scalability issues.

In simple terms this means that very few products are able
to:

- provide two-way communication with existing databases
  in real time and with the guaranteed maintenance of
  data integrity

- access or provide servers for GI over existing networks

- offer an ability to expand over time – scalability

- operate on existing equipment and within existing
  operating environments.

By design, compatibility and scalability are less of a problem
within open system environments. User choice will
progressively improve as more suppliers recognize that
growth in GIS markets will come from a massive latent
demand for large-scale, distributed solutions that are an
integral part of IS infrastructure, rather than the separate,
small-scale, local solutions that the industry has relied on.

Open IS standards in the form of EPHOS and Posix are already in place to ensure that many forms of IS can be a part of an integrated IS infrastructure, but extensions to these standards and some new standards are required if GIS is to become an integral part of such environments.

3.2.2   Application portability

Application portability is another area in which GIS users and suppliers could benefit from open standards. Application development represents a major expense and risk, and any initiatives which encourage choice, economy and reduction of risk will be very attractive.

For the concept of application portability to become a reality, the GIS industry will require standards on a variety of areas including GIS functionality, Application Programming Interface, Macro Language, User Interface and internal data models. Achieving this will inevitably be a long-term process, but even some degree of standardization would improve the portability of application software, improve choice, and reduce costs and risks.

Ultimately the portability of GIS-based applications would encourage the development of a packaged-application market. Although some GIS-based application packages are available, they are dependent on proprietary products, that is they are non-portable applications. The wider availability of packaged application software would substantially reduce the costs and risks currently associated with application development.

Application portability is also good news for GIS suppliers since emerging markets for packaged applications have the potential to be more profitable than those which already exist for providing bespoke implementation services.

3.2.3   System usability

The advantages of usability of IS products have been amply demonstrated in recent years and the strong trend towards Graphical User Interfaces (GUIs) is recognition that usability is now taken very seriously within most parts of the IS industry.

GIS are graphical by nature, and the vast majority of products have supported concepts such as windows, icons, menus and pointers for some time. However, it is generally true that GIS-based applications remain more complicated and harder to use than many other forms of IS, and this is an area where future progress can be made.

Standardizing the user interface is already happening with the emergence of Motif for workstations and Microsoft Windows for PCs. Further standardization would probably represent interference with industry opportunities to innovate and differentiate products – analogous to standardizing the layout of hi-fi controls, for example; but consultation between users and suppliers could help establish a greater degree of consistency and promote the use of best practice. Commonly used concepts, such as information layers and gazetteers could be extended, and GIS terminology could be used more consistently so that users familiar with one system can quickly adapt to another.

3.2.4 Interoperability/ interworking	Interoperability is one of the key GIS standards requirements. As the extent of integrated IS infrastructures grows it will become increasingly important for GIS products to communicate and interwork with other systems in client/server type arrangements. Such interworking will be required within distributed organizations and between several organizations. Once such interoperability is standardized, it will make it possible for Value Added Network Services (VANS) to be provided. These will allow direct on-line access to many sorts of GI, but will be particularly suitable for accessing time-critical datasets such as land registration details, weather, river levels or traffic status.

Projects such as Domesday 2000, National Land Information Systems (NLIS) and Northern Ireland Geographic Information System (NIGIS) will use EDI to provide direct on-line access to GI between participating organizations over wide-area networks. EU funded research is being undertaken by Data Centralen in Denmark to create a multinational environmental information network.

An effective standard for GIS interworking will also help individual organizations – particularly those that require several different GIS products to serve their requirements. For example, some organizations, such as National Rivers Authority, have very diverse requirements including real-time data capture, satellite imagery, cartography and simple data query facilities, and these requirements may lead them to implement an infrastructure containing several GIS products. Within such an environment, individual products will need to be able to exchange on-line data and peripherals in a distributed client/server configuration. However, there is unlikely to be a single standard covering all aspects necessary to achieve full GIS interworking, and the best hope for the GIS industry is to 'piggy back' on relevant emerging or accepted standards, de jure or de facto, such as EDI-based standards.

Several leading GIS vendors now offer PC-based Data Acquisition and Report Tools (DARTs) with a capability to provide simple and economical access to data held on more expensive and sophisticated GIS products.

Generic Electronic Data Interchange (EDI)-based standards can also make interworking viable for GI/GIS, but further standards work is required to tackle issues concerning the packaging of GI for transmission over networks, and there needs to be much greater understanding in the GIS community of issues such as ownership, charging, privacy, liability and copyright.

EDI and open systems issues are discussed more fully in Chapter 4.

3.3	**Problem areas for de jure GI/GIS standards**	Whilst the arguments in favour of de jure GI/GIS standards are theoretically very sound, in reality the case is undermined by a range of factors.
3.3.1	Awareness and commitment	One of the main constraints on standards development is that standards are fully understood by relatively few. A significant proportion of IS standards initiatives are instigated by rare individuals who have a clear perspective of the future of the IS industry and an adequate appreciation of open standards opportunities. In general, those who make strategic IS decisions or are responsible for the technical direction of products or services are more likely to appreciate the value of standards than most others.
3.3.2	Scope of influence	One problem with the standards agenda is that some potential areas of standards activity may be outside the scope

of influence of standards organizations. For example, because the markets for GIS are international, British IS standards are unlikely to influence the development of products from North America and Australia.

3.3.3 Participation (of the right people, at the right time)

At the different stages of standards-making, different sets of people are often needed to contribute differing expertise in technical know-how, user needs, and the standards-making process itself. Often they do not come forward. Sometimes they are unaware of the need for open standards; often they are not convinced there is enough in it for them, given the high cost and slow progress of most standards making.

For a de jure standard to be widely acceptable, it must be seen that its development was not dominated by an individual supplier or user, and the standards must be at least as effective as current best practice. Standards therefore require participation of a representative sample including major users and major suppliers, and ideally should be overseen, managed and championed by an impartial individual.

In *User Needs in Information Technology Standards*, Dr Harald Nottebohm suggests that:

> *Suitable techniques of co-ordination, interviewing, etc, must be used to motivate experts to contribute their knowledge without having to write something, speak an unfamiliar language or spend time on meetings if they don't like it or are not allowed to.*

3.3.4 Funding/non-specific business benefits

Probably the single biggest problem for those involved in the development of GI/GIS standards has been in obtaining funds. Whilst sometimes those who participate in regular standards meetings do so at their own expense, some activities inevitably incur costs which it is unreasonable for individual people or organizations to bear. Such costs are usually incurred through administration, management, standards drafting and attending meetings in the UK and overseas.

The AGI standards scoping and strategy report identified the need for a much greater level of resourcing in terms of participation by various organizations.

The AGI has had some success in finding sponsors for some standards projects, particularly for standards that have had a direct relevance to the sponsor. Notable examples have included the development of BS 7666, where the LGMB has co-ordinated and largely funded the work. Whilst there is scope for extending initiatives of this sort, AGI has had limited success to date in securing support for the development of more infrastructural standards where the immediate benefits are less obvious.

Central to this problem is that both public and private sector management are expected to look for a visible return on any investments they make, and when it comes to contributing to long-term initiatives for the common good, the investment will inevitably seem a lot less compelling. This is exacerbated by the fact that the GIS industry contains a few very large companies and many small ones; consequently the ability to contribute and the case for participation is by no means universal.

3.3.5 Timescales

By its nature standards-making involves considerable consultation, and often this requires large numbers of individuals from many separate organizations. This consultation process usually involves a substantial quantity of correspondence and is vulnerable to the disagreements which can cause long delays. In *Information Technology Standardization: theory, process and organizations*, C F Cargill points out that:

> The consultation period is dependent on the number of participants involved, and some organizations find this feature of the standards making process easy to exploit for their own ends.

It is generally very difficult to estimate the timescales and resources needed for the development and adoption of a standard. Although some standards organizations have procedures for 'fast-tracking' some items, standards take several years to be agreed and there is a further delay before industry implements them. In the meantime market forces will often be at work developing de facto standards and thereby diminishing the useful effects of de jure standards.

The ability of the market to adopt new technology is now much shorter than the development time for de jure standards, and de jure standards could be said to have a window of opportunity within which to make an impact. Aside from the difficulties this causes in terms of activity planning and resourcing, it also creates pressure for standards developers to anticipate future needs. Anticipatory standards are harder to justify because they often have to be initiated in the absence of significant demand, and they are more difficult to specify because of an absence of previous experience.

3.3.6 Planning and prioritization

Another problem with developing GI/GIS standards is that planning and prioritization are very difficult to get right. There are a two main reasons why this is the case: organization-specific priorities and inconsistent funding.

Organization-specific priorities

Individual organizations are at very different stages with GIS projects and therefore have very different priorities. Those few organizations that are attempting to implement corporate GIS, for example, are more likely to appreciate the importance of interoperability than those implementing personal GIS solutions.

Inconsistent funding

Due to the above organization-specific priorities, funding for standards initiatives may not follow the logical order in which things ought to be done. In particular, application-specific standards may proceed much faster than generic standards on which they should be based.

3.3.7 Implementation (by users/suppliers)

Although the development and formal adoption of a standard is a significant achievement in itself, it does not necessarily follow that the standard will be used in practice. Whilst GI/GIS standards will have the long term effect of reducing costs, in the short term additional costs may be associated with their implementation.

In the case of GI/GIS suppliers, some may be required to undertake a major redesign and redevelopment of their products at significant cost; and although users appreciate the benefits of GI/GIS standards, some may take the view that specifying a standard whilst it is still new may constrain choice. In short, few organizations will implement a new standard until it is known to be stable and effective. DIGEST and IHO/S57 have announced that after the currently pending versions are issued, there will be a two-year moratorium on change, to allow vendors to implement the standards.

3.4 The need for greater support

Whilst this publication gives much attention to the opportunities which GI/GIS standards provide (whether these are de facto or de jure standards), it is useful to consider briefly what is likely to occur if members of the GIS community fail to support the implementation of standards within their respective areas.

For geographic information systems suppliers

Current sales of GIS are very small by comparison with the latent demand for GIS. Recent estimates suggest that the UK market for GIS is around £57 million, only part of which is spent on products and implementation services. Anecdotal evidence suggests the latent demand for GIS is somewhere between 10 and 100 times this figure.

Market research confirms that most GIS sales have been for personal and small departmental systems, with 85 per cent of systems supporting five users or less. This is consistent with the view that GIS is still very much regarded as a separate specialist technology – far from the concept that most GIS suppliers promote.

For the full potential of the market to be realized it is necessary for GIS to develop its role and become more acceptable as an integral part of IS infrastructure. This will inevitably require a much greater level of compatibility, portability and scalability of solutions than is evident in the current GIS market, and it is unlikely that this could ever be fully or properly achieved without the presence of further standards.

To be part of this larger, more integrated GIS market, suppliers may choose to bear the costs and risks of achieving compatibility, portability and scalability alone. Certainly some suppliers are much better placed than others to achieve this.

A more economical and lasting approach would be to contribute to the development of open GIS standards by working together for standards in non-competitive areas. An example of this type of approach is the way that the petroleum industry has collaborated with the Petrotechnical Open Software Corporation (POSC), see subsection 4.1.6.

In summary, the benefits will be:

- reduction in number of disparate standards and formats which have to be supported

- opportunity to develop the market (for those able to make the transition between a low-volume high-cost market to a high-volume low-cost market).

For suppliers of geographic information

The relationship between the market for GI and the market for GIS is not unlike the relationship between the music and the audio industry – neither would survive without the other. The credibility and viability of GIS as a useful business tool is strongly associated with the suitability and cost of available GI.

The GI industry needs to be able to demonstrate that it is competent at handling GI so that its customers have confidence in information derived from its services and products.

Current players in the GI industry will face competition from newcomers, many of which will seek to compete on price and speed of service. Without recognized standards for GI, many established and experienced companies may be unable to differentiate their products and services from inferior ones.

The presence of GI standards would enable the industry to gain recognition for the added value that good quality products and services provide, and in time it could create a market for standard data products and services several times larger than existing markets for GI in proprietary formats.

In summary, the benefits will include:

- reduction in the number of competing data standards

- the ability to create markets through the use of 'standard' data products. An example of this is the investment currently being made in the development of Graphics Data File (GDF) road networks for Europe.

For users

For GIS users, the absence of standards applying to their particular applications may or may not be a problem depending on individual circumstances. For many applications it may not be necessary to adopt a formal standard at all, but simply to put in place agreements on specific issues instead, and create an internal corporate standard.

The need for standards will arise particularly within projects that have a number of participants. For example, analysis of international air pollution would require some degree of standardization to ensure that data collected by participating countries could be presented and used objectively.

In the absence of suitable standards, users may be unable to use data provided to them, they may incur additional costs in converting data into a consistent form, and perhaps even worse, they may be completely misled by subsequent analysis and use.

3.5   **Benefits of**   In summary, the benefits of standards will vary depending
       standards   on the organization, but can generally be expected to be:

- reduced cost of investment and technical risk of implementing GIS

- the opportunity to implement GIS more widely as a generic information processing tool within an organization

- benefits through seamless integration of spatial and non-spatial information systems

- portability of applications and of staff across applications with a reduced learning curve

- a reduction in overall maintenance and running costs.

4   Status of GI and GIS standards

4.1   Organizations involved in GI and GIS standards

GI/GIS related standards bodies are either area-based or interest-based; they have representatives from organizations based on either geographic proximity or common professional interests, respectively. These groups are increasingly working together to establish general purpose standards, as far as possible.

4.1.1   The United Kingdom

Within the UK, the main bodies are as listed below:

- the Association for Geographic Information (AGI)/British Standards Institution (BSI) joint standards committee, IST/36

- the Local Government Management Board

- the National Joint Utilities Group.

AGI and BSI IST/36

IST/36 is a technical committee of the British Standards Institution with representatives from a wide range of parties, including end-user representatives, vendors and data providers. It is a joint committee with the Association for Geographic Information (AGI) standards committee. IST/36 has commissioned and approved several British standards, related to geographic data transfer, which BSI publish.

BSI is concerned only with de jure standards, whereas the AGI standards committee is also concerned with de facto standards. Whilst there is a widespread view that the AGI standards committee work has focused on de jure standards, of the 36 standards-related projects defined in the AGI standards strategy, 15 are concerned with de facto standards only, 8 with de jure standards, and 13 contain some element of both de jure and de facto standards.

LGMB

The Local Government Management Board has worked closely with the BSI IST/36 committee and the AGI to develop standards of benefit to local government and others. The LGMB contribution in the last two to three years has been pivotal in the finalization of BS 7567 and the development BS 7666.

NJUG	The National Joint Utilities Group has undergone some changes due to the privatization of the utilities industries. It is now focusing on common non-competitive requirements and will develop a more detailed work plan for GI standards in 1994/95.

4.1.2 Europe

Several European countries standards organizations (for example AFNOR and DIN) have developed GI/GIS standards, again mostly concentrating on data exchange standards. These bodies, along with BSI, are now represented by a CEN technical committee TC/287. The work programme of CEN TC/287 includes development of a general reference model for geographic information standards and the development of specific standards for positional referencing and spatial data transfer. Details of the standards agenda for committee TC/287 are included in Annex B, and a table comparing the UK and European agendas for standards is included in Annex C.

GI/GIS related standards work is also being undertaken by CEN TC/278 which deals with Advanced Transport Telematics, including road maps for in-car navigation.

4.1.3 North American countries

There has also been considerable investment in GI-related standards in the USA and Canada. The USA National Institute for Standards and Technology (NIST) has produced a Federal Information Processing Standard (FIPS) for GI transfer called SDTS. This is mandatory for all Federal Government departments. Canada has produced a set of standards for data transfer called SAIF.

4.1.4 International organizations

A new ISO committee, ISO TC 211, entitled 'Geographic Information (Geomatics)' was announced in June 1994. The scope of this committee is defined as geographic information and applications of geographic information. The Norwegians are providing the secretariat and the first meeting will be in Oslo in November 1994.

The original Canadian proposal was for a committee to progress standards on 'Geomatics', which was defined as a generic term covering diverse aspects as geodesy, cartography, toponomy, geographic database management, as well as the data exchange of geographic information. However, the title and scope were modified in the considerable discussion and debate that took place prior to the committee being set up.

4.1.5   Sector-specific organizations

There have also been several application-driven standards activities:

- the Digital Geographic Information Working Group (DGIWG)

- the International Cartographic Association

- CEN TC/278

- the Comité Européen des Responsables de la Cartographie Officielle (CERCO)

- the International Hydrographic Organization.

Digital Geographic Information Working Group (DGIWG)

The development of DIGEST has emerged from requirements for common data transfer of military maps. DIGEST has been developed as an international collaboration of the major world defence mapping agencies through the DGIWG. It is a NATO standard (STANAG).

International Cartographic Association

The International Cartographic Association has a commission on geographic data transfer, and has produced a number of documents on the subject, but the work of the ICA has not yet developed into formal standards.

CEN TC/278

This CEN committee is responsible for GDF, the Graphics Data File, which has been developed primarily for in-car navigation systems by the major car manufacturers. This is currently being put forward as a draft European standard (EN) and is being submitted to the equivalent ISO 204 committee on Transport Telematics.

Comité Européen des Responsables de la Cartographie Officielle (CERCO)

A European Territorial Data Base (ETDB) project has produced a catalogue of data classes and features. The project is funded by CERCO (the heads of the European mapping agencies) under their MEGRIN initiative.

The International
Hydrographic
Organization

The International Hydrographic Organization has published an Object Catalogue standard S57 (formerly special product SP57) and Transfer Format (DX90) for vector hydrographic information.

4.1.6   Other organizations

Outside the de jure initiatives there are two initiatives aimed at influencing the GIS vendors directly through pressure from GIS users:

- the Open Geographic Information Systems Foundation (OGISF)

- the Petrotechnical Open Software Corporation (POSC).

POSC and OGISF remain the only major groups currently attempting to influence GIS vendors and their products as well as developing data exchange oriented formats.

Open Geographic
Information Systems
Foundation (OGISF)

The Open Geographic Information Systems Foundation is a new forum which is emerging from the Open GRASS Foundation. See subsection 4.3.3.

Petrotechnical Open
Software Corporation
(POSC)

POSC is an international non-profit organization founded in October 1990 with a mission to define, develop and deliver Open Systems standards and software integration platforms for the Exploration & Petroleum (E&P) industries. Membership now stands at around 75 organizations, and POSC has a staff of around 50 people with offices in Houston, Texas and London. Funded and staffed to provide pragmatic and timely solutions to common industry computing and data problems, POSC also provides open specifications complemented by sample implementations of the specifications. POSC will also brand commercial products for compliance with its specifications.

While the trust is systems-related in general, the largest effort of POSC has been addressing data and its management. POSC's results are aimed to provide the E&P industry with a means to achieve a higher level of flexibility – to allow a company to react faster to changing market conditions, reallocating resources where needed and taking advantage of new technology sooner, by eliminating the burdens of rigid computer systems.

POSC uses RFT (Request for Technology) and RFC (Request for Comments) procedures (analogous to those used by the Open Systems Foundation) as powerful tools to harvest state of the art technologies available from the market place.

POSC solutions include:

- the endorsement of base computer standards that facilitate integration

- the Epicentre data model which provides common reference to all E&P disciplines

- the POSC data access and exchange specification which provides common access to data across diverse database management systems

- the POSC exchange format for loose integration of application and heterogeneous data

- a user interface style guide which defines a consistent 'look and feel' for software from different vendors.

POSC is also addressing:

- technology transfer of its work to the E&P industry through training, migration consultancy, guides and software demonstrator tools

- ensuring that technology works appropriately through the services of an interoperability lab (ILAB) to ensure that the systems work in practice, and to assist with testing of products

- migration issues through a special interest group looking at common issues relating to POSC standards.

POSC is also involved in a number of de facto and de jure standards bodies including:

- the CAD/CAM based STEP group

- ISO SQL3 committee

- OSF

- X/Open

- Computer Graphic Metafile*petroleum industry profile (CGM*PIP) initiative.

The issues that POSC raises which are most relevant to GIS standardization are:

- the benefits of standards have been recognized by the E&P industry which is willing to fund an international 50-strong organization to act as an enabler for the introduction of appropriate standards

- the approach is user-led but with full vendor involvement

- POSC intends to brand products in addition to producing specifications

- POSC participates in both de jure and de facto groups depending on which will get the desired results

- POSC may well be able to provide a model for GIS standardization work

- there is an overlap between E&P general standards requirements and GIS standards requirements. There is scope for direct co-operation between POSC and GIS standards bodies.

4.2   Existing standards

A large number of data-related standards have been developed within certain countries or specific application areas, mainly to meet the need for geographic data exchange.

This section covers data and related standards which are in a more advanced state of development than system-related standards. For a discussion of system-related standards, see subsection 4.3.2.

In general, data-related standards associated with geographical information concern:

- spatial units

- data quality standards

- transfer mechanisms

- standard data models.

Spatial units

These standards will identify a definitive set of basic spatial units (BSUs), identify their owners and the community of users for this data, and consider issues relating to referencing locations.

Data quality standards

Standardization in this area includes the production of appropriate generic quality assurance and control definitions, guidelines and indices.

Transfer mechanisms

In the UK this area of standardization has primarily focused on the development of the NTF standard which has recently been published as BS 7567.

Standard data models

Standardization here includes the street gazetteer, land and property gazetteer (projects which were inspired and have been led by the LGMB). These are now BS 7666 (see below).

4.2.1 Data transfer standards

Within the UK, the main de jure GI-related standard developed to date is BS 7567 (Electronic Transfer of Geographic Data) which was developed from the National Transfer Format (NTF), created by the Ordnance Survey, AGI and a wide variety of user input from land and sea mapping agencies, local authorities and utilities.

It is less clear what the main de facto standards are, as there is no independent body such as X/Open to identify and agree the industry standards. However, it is clear that within the UK, the best established de facto standard is DXF (which is 'borrowed' from the CAD market), followed (in no particular order) by ArcExport (ESRI), and SIF (Intergraph Corporation) for vector data and possibly TIFF for raster data (though there is a plethora of raster formats). Within the UK, Ordnance Survey recognize the need to support de facto as well as de jure standards through making mapping data available in both BS 7567 and DXF formats.

Data transfer standards must be derived from a data model and so are unlikely to become de jure in advance of data model standards being agreed. CEN has given priority to work on the standards framework, definition of terms and data models as essential prerequisite standards for data transfer and others.

Related standards

There are a vast number of existing and developing standards which can be used for transferring geographic data, though some have been developed for specific purposes.

In addition to BS 7567 and DXF these include (this list is not exhaustive):

ASCII – a 7-bit character coding convention
ATKIS – German national format
DIGEST – developed by NATO for defence mapping agencies
EDIGEO – French national standard
ETAK – used for road network maps
GDF – used for in-car navigation and related purposes
IGES – used for CAD
IHO S57/DX90 – used for maritime mapping
PDES – one of the CAD-related STEP standards
SAIF – Canadian national format
SDTS – the US federal government standard
TIGER – used by US government and census bureau.

In addition, there are other general standards which could be used including:

CGM – Computer Graphics Metafile, part of an ISO standard
EDIFACT
HPGL – Hewlett Packard Graphics Language
MAP/TOP
PHIGS – an ISO standard for graphics.

A full examination of the relationship between these standards is beyond the scope of this volume. This volume focuses on some of the main standards below.

11-layer model

It has been proposed that all GI data exchange standards should be related to an 11-layer model, analogous to the OSI 7-layer model. This model comprises the following layers:

- media, the physical medium used

- encapsulation, the encoding of the message in a physical format, such as ISO 8211, EDIFACT

- world view, the agreement on features and relationships through a common conceptual data model

- data structure, the logical data model

- schema implementation, the physical data model

- feature/attribute, the detailed catalogue of how both of these are encoded

- content, the description of what is contained from the provider's view

- metadata, the quality, source, history, extent of data

- directories and indices, the retrieval mechanisms to be used

- receiver tools, any agreements covering tools to be used for visualizing the data, or importing the data to a GIS, agreements covering data or co-ordinate transformations

- environment, agreement on software environment where necessary.

The 11-layer model is described in more detail in the volume: *Some Questions and Answers About Digital Geographic Information Exchange Standards* by C A Kottman.

Main existing standards

Some of the main existing standards include those listed below:

- DIGEST (DIgital Geographic Exchange STandard)

- Spatial Data Transfer Standard (SDTS)

- EDI and EDIFACT.

DIGEST

DIGEST is actually a family of standards containing three separate standards.

DIGEST-A

DIGEST-A is:

- a feature based standard

- a transfer format analogous to BS 7567

- intended to be used for exchanging GI data using transportable media.

DIGEST-B

DIGEST-B is:

- the same as DIGEST-A, except for using ISO 8824 for the encapsulation

- intended to be used for exchanging GI data using telecommunications

- the least important of the three standards.

DIGEST-C

DIGEST-C is:

- a relationally based standard

- a direct use data standard

- also referred to as:
  - VPF (Vector Product Format)
  - VRF (Vector Relational Format)
  - MIL-STD 600006

- the format in which the Digital Chart of the World data (DCW) can be input into GIS applications using DIGEST interpreters developed by a number of GIS vendors.

DIGEST was developed by DGIWG (Digital Geographic Information Working Group), an ad hoc group of NATO military defence mapping agencies. Version 1 was released in

1991, and Version 2 in 1994. Version 2 includes compatibility with IHO (International Hydrographic Organization) standards. DIGEST will be one of the standards submitted to the new ISO committee when it is formed.

Spatial Data Transfer Standard (SDTS)

SDTS has been ratified by NIST as a Federal Information Processing Standard (FIPS 173). It is sponsored by the US Federal Government Geographic Data Committee.

This committee has until recently been chaired by the head of the US Geographical Survey (USGS) but it is now chaired by one of the senior ministers in the US Federal Government, due to the emphasis placed on the development of IT and 'information super-highways' by Vice President Al Gore.

The USGS and the US Census Bureau are committed to using a subset of this standard known as Topological Vector Profile (TVP) for future releases of Census and USGS data. Other Federal Government organizations will also be encouraged to use TVP, although this is not mandatory for local government in the US. SDTS has also been adopted by the Australian standards committee as the national GI data exchange format.

EDI and EDIFACT

EDI standards can be used for encapsulation of geographic data and are being used in Finland for requesting and receiving geographic information updates from the national mapping agency. EDI has already gained widespread international acceptance for commercial applications in a wide number of market sectors. It is beginning to be used in European projects which involve exchanging data, such as EDICITIES which is an experimental project linking three local authorities in Holland, Italy and France, and in the EU-sponsored DRIVE programme for the exchange of geographically referenced traffic related information, such as major road incidents.

GIS raises many issues of information management and there is a potential overlap with document exchange standards. These are related closely to the 11-layer model discussed above and cover issues such as structure, interchange, physical media, data transmission, image formats, multi-media and overall document structures. As with the GIS market, there are a large number of competing standards and it is too early to be sure which will emerge as the market leaders.

4.3   Ongoing standards initiatives

The main initiatives can be grouped into data-related, systems-related, the Open GIS Foundation and underlying general initiatives.

4.3.1   Data-related initiatives

Within CEN TC/287 and the IST/36 standards work, the following areas have been identified as requiring further development. A table comparing UK and European standards agendas is included in Annex C.

AGI/BSI IST/36

The concerns of the joint AGI/BSI IST/36 committee cover:

- spatial unit definitions

- spatial unit boundary data

- spatial unit attribute data

- spatial referencing

- quality model

- quality assessment of data

- quality assurance

- spatial data types and standard data models

- street gazetteer

- land and property gazetteer

- address standard

- conceptual data models.

Spatial unit definitions

This aims to provide a register of sets and hierarchies of spatial units which subdivide the UK for a variety of administrative purposes.

Spatial unit boundary data	This is a similar project to register possibly competing boundary data sets which meet appropriate quality criteria.
Spatial unit attribute data	This is a complementary register project to record data listed by reference to registered spatial units; a start has been made on this through the compilation of a list of UK government datasets.
Spatial referencing	Indexing methods for a variety of large datasets that will continue to be stored independently of their location within the cartographic projection.
Quality model	This is a glossary of the characteristics that are used to express the quality of a dataset; for example, completeness, accuracy.
Quality assessment of data	This project, in two parts, is aimed at setting standards for the quality of data; the first part is a specification that will be mainly concerned with establishing standard accuracies and interpretations of the real world; the second part is aimed at quality control procedures.
Quality assurance	In order to obtain consistent quality, a quality scheme is necessary. This is covered by BS 5750 (ISO 9000) but it is generally felt that, for geographic data, individual organizations have too much freedom to choose an ineffective implementation.
Spatial data types and standard data models	These projects aim to provide sufficient standardization of data for open systems to work effectively.
Street gazetteer	This is a key step in implementing recent legislation to improve control of street works; an unambiguous index of streets, their location and extent is fundamental to this exercise. This is BS 7666 part 1.
Land and property gazetteer	This is related to the Street Gazetteer via a property address element; it is intended that every separately identifiable parcel of land/property will be referenced in the gazetteer; it is anticipated that some form of legislation will be required to enforce collection of the data. This is BS 7666 part 2.

Address standard	This will provide a common format for addresses (unique identification and location), within Great Britain, and is BS 7666 part 3.
Conceptual data models	This is based on a review of the NTF data model described in BS 7567, other transfer standards and proposals and the Ordnance Survey Conceptual Data Model 3.1.
CEN TC/287	Aside from other CEN initiatives dealing with data quality, coordinate systems and metadata, the concerns of the CEN TC/287 committee cover:

- development of a reference model for GI standards

- development of a generic data model

- development of a European transfer format.

Development of a reference model for GI standards	This is planned to be the first deliverable from CEN. It will identify the requirements for specific European standards and provide a framework for these and related standards (it should be noted that once a draft standard is presented there is a lengthy formal procedure to be followed).
Development of a generic data model	This will describe the basic elements which are used to describe features that exist in the real world. It is due to be produced as a draft CEN standard in 1995.
Development of a European transfer format	Based on existing transfer formats, this is due to issued by CEN TC/287 as a draft standard by 1995.

4.3.2 Systems-related initiatives

The goal for GIS standards is to achieve coexistence, scalability, interoperability and integration with other IT systems within an organization. This must include a two-way exchange of data with other corporate IT systems. In particular, non-GIS must be able to access the geographical data in a GIS where the output is, for example, in alphanumeric list form rather than graphical or through a basic graphical data viewing facility.

Standards required	In order to bring about the above, standards are required in the following areas:

- applications development methods, tools and languages, for example methods such as SSADM and CASE tools which support GIS analysis and design, standard spatial operators to enhance SQL. This may lead to the development of a standardized Applications Programming Interface (API) for GIS

- human/computer interaction. This would include a standard GIS user interface as an extension to OSF/Motif or similar standards, again possibly through appropriate APIs

- data interchange and access, based on SQL, IRDS and similar standards. This will include the definition of spatial operators to handle spatial objects

- systems management and administration. The necessity to handle large quantities of geographical data, possibly in several different formats, will generate system management problems that will need to be considered in the Posix work on system management standards and similar

- guidelines for the 'soft issues' involved in GIS implementation. These include guidelines on project management and standardized cost/benefit analyses. The aim of these guidelines is to assist and encourage broader use of geographic information. It is unlikely that they will become standards in their own right, but they will make extensive reference to existing and developing standards

- benchmarking standards. This will provide GIS price/performance measurements based on a suite of programs testing core functions, analogous to the benchmarking work undertaken in the IT world by groups such as the Transaction Processing Council

- conformance testing. Once standards are defined and widely accepted, it is still essential to ensure that conformance to published standards can be authoritatively tested.

AGI standards work

Work already undertaken in this area by the AGI standards committee includes:

- development of terms of reference for GIS implementation in a de facto/de jure standards based environment

- significant contribution to the development of the SQL3 standard by the AGI. The aim of this is to ensure that SQL3 will be better able to support the needs of geographic information management

- initial proposals for a standard GIS user interface and input to the ISO GUI standards committees

- input to the BSI IST/20 IRDS standards committee to discuss the way in which geographic information will be described, handled and stored

- the conceptual data models within the data transfer standard and significant contributions to the review of ISO 8211, a data descriptive file for information interchange

- a directory of existing and proposed standards (UK only). This document provides a safeguard against duplication of effort on creating new standards where others already exist. AGI has a directory which is currently undergoing review.

The emphasis of the CEN committee is on data-related rather than systems standards and this reflects the emphasis of many other standards committees world-wide. This is due to the fact that geographical data exchange is a fundamental requirement for almost all GI and GIS users and therefore this has had to be the first problem to be resolved.

Another major initiative that will emerge in 1994/95 is the introduction of an ISO committee on GI and GIS related issues. The structure and scope of this committee are still under debate, as well as its relationship with CEN TC/287 and how the Vienna Agreement (which governs the relationship between ISO and CEN) will apply.

4.3.3 Open GIS
Foundation

Within the US, an organization has recently been formed to address issues related to open systems GIS. Called the Open GIS Foundation (OGISF), it aims to replicate the successful work of groups such as the X Windows consortium and OSF in producing the widely accepted Motif standard. It also aims to produce an operational model called OpenGIS which will be a public domain software specification to enable interoperability between diverse software systems and data structures.

The OpenGIS operational model incorporates a virtual data model and an Application Programming Interface (API) aimed at providing transparent access to spatial databases from any application. The aim of this OGISF work is to provide a common base for developing tools for application developers to locate and use any geographical data available on a network. This will mean that vendors will be forced to compete on a more even playing field and this will help develop the market for GIS.

OGISF has already gained support from, amongst others:

- the Open GRASS Foundation
- US Army Construction Engineering
- US Department of Agriculture
- MIT
- Intergraph Corporation, Federal Systems Division
- NYNEX
- State of California.

4.3.4 General initiatives

There are also concepts and definitions that underlie any successful GIS implementation. The aims of activity which addresses these issues are to provide:

- a common language for those involved with GIS as educators, implementors, data or system providers; there is scope for definitions to be included in relevant published standards, as appropriate

- definitions which should be used as a basis for all standards, for example, common functional and data models (which would help define the core functions of a GIS, either in general or for a particular vertical market sector). This also includes identification and description of each of these vertical market sectors.

Examples of work undertaken by the AGI, in each area respectively, include:

- a dictionary of geographic information terms (UK only). This comprises a list of terms used for geographic information with explanation and is available as an AGI publication. Such work is considered appropriate by CEN for European standardization as it is the basis of a common understanding

- a geographic information glossary (UK only). This is intended to expand on the more complex terms in the dictionary with, if necessary, the aid of diagrams and examples. A typical example of a term which merits this treatment is 'topological'.

4.4 Guidance on using GI and GIS standards

At the current time there is no clear evidence to indicate which standards have a long-term future; there is no clear evidence of de facto standards emerging, and the few de jure standards that have been developed in the UK are very new and largely unproven. Consequently, until the situation becomes very much clearer, it is recommended that organizations adopt a defensive approach to GI/GIS procurement and implementation.

While this will be dependent on the nature of the organization, some general principles for protecting long-term investments in GI and GIS include:

- apply open standards whenever possible, for example, GOSIP or Posix

- specify software that will run on a variety of hardware platforms; ideally on standard or open standard hardware platforms or on a platform that is already in common use within the organization

• specify systems with an inherent ability to expand to meet foreseeable needs; ideally define outline GIS requirements within the GIS/IS strategy for the organization, and select a system that can best accommodate long-term needs

• insist on an ability to import and export data in an ASCII format of some description, and ask for a complete definition of the formats

• ensure that the system can support NTF for data input; preference should be given to systems that can export NTF data.

• define corporate standards for GIS-related initiatives; particularly where the intention is to build an integrated GIS/IS infrastructure.

• specify systems which offer other useful data interpretation facilities

• bear in mind that the most expensive and lasting investment will be the data itself; look for other ways in which to protect this investment such as update commitments from data suppliers.

An ASCII format will provide a good chance of future interpretation, but be warned that it is important that datafiles contain the full data model, rather than just the graphics. DXF is an ASCII format, but it is deficient in a number of important respects when used for the transfer of geographic data. In particular it does not support the full reconstruction of the data model.

Main drivers

One of the main drivers for open systems has been the specification of open systems requirements in tenders from the EU, the US and other governments, and major organizations such as utilities. In accordance with EU directives, local government procurement also requires adherence to open systems standards. This will therefore have a major impact on all GIS vendors whose primary markets to date have been local and central government and utilities.

Development of a strategy

A further issue is that whilst many organizations find that the development of a corporate IS/GIS strategy is essential, it is a time consuming process. It is important to ensure that a sound business case can be made for GIS, that full advantage is taken of any existing systems, and that new and potentially important uses of GIS are identified and developed. An organizational IS strategy facilitates adherence to these good practices.

Interim solutions

However, specific business units often wish to start their own initiatives in advance of the corporate strategy. By adopting suitable standards for interim systems, organizations can protect the opportunity for these systems to be extended in the future. This has major benefits of being able to grow systems without major rewrites and changes of direction as the system develops, provided an initial level of analysis has been undertaken. It can also allow earlier (and therefore cheaper) data capture in important areas, which will support better working for the individual business units, and faster implementation of the corporate system while minimizing duplication of effort.

Some of these benefits would be available using a corporate strategy based around a proprietary GIS solution. In the long run, however, open systems GIS will provide cheaper solutions, and will support closer integration with other IT investments.

There is some concern in the industry that as many businesses fragment into self-managed business units (this trend is affecting central and local government and the utilities as well as the traditional private sector) the corporate GIS strategy becomes less of a practical reality and more unachievable pie in the sky. Yet there is likely to be a continuing need, in many cases, for the exchange of data and interworking of systems across these new organizational boundaries. In an environment of departmental or corporate systems, standards become **essential**.

5   Likely future developments in GI and GIS standards

5.1   Vision of the future GI/GIS markets

To appreciate how future GI/GIS markets could be impacted with the introduction of appropriate standards, it is useful to take a considered view of the future of other parts of the IT industry. The trends in IT over the next few years are likely to include:

- continued advances in the performance of systems

- development of more reusable software and data, especially based around object-oriented paradigms

- greater use of network services based around ISDN and OSI

- the development of distributed computing environments, especially client/server configurations, based on OSFs DCE and Unix International's Atlas

- increasing sophistication of GUIs and general Human Computer Interfaces

- the battle for desktop computing

- better visualization and the use of multimedia

- greater awareness of information management issues.

All of these factors will have a significant impact on GIS markets by presenting many new opportunities. If the GI/GIS community is able to implement appropriate standards users and suppliers will be able to make full use of the opportunities presented.

Development problems

There is frequently a significant amount of work involved in bespoke development to provide a working GIS. The development of more reusable code would help reduce this effort to some extent and would encourage the availability of packaged applications. If appropriate object-oriented standards are agreed and widely adopted, then a more document-centred approach will, for example, allow the use of embedded geographical objects within other systems.

Network services
: The availability of broadband networks and distributed computing environments would allow greater integration of systems across geographically diverse organizations and more points from which users could access GI.

Query language
: The establishment of a 'Geographic Query Language' based on SQL, the use of client/server models and the development of third-party GUIs will affect the architecture of GIS products. This will mean that, in the future, the emphasis for GIS development will focus on developing GIS analytical functions, and this will generate significant re-engineering of GIS packages/products as they currently stand.

Desktop systems
: One major area of impact would be in desktop systems. IT hardware vendors are already aware that the markets for proprietary solutions in the mainframe and mini areas are dwindling and that this will continue during the 1990s. The growing market will be for desktop based systems and competition is already developing rapidly in these areas.

Multimedia
: The impact of these forces will be the provision of low cost, highly functional, multimedia, GUI based systems which will be able to provide an excellent delivery mechanism for a wide variety of applications including GIS. High volume GIS sales will depend on the GIS industry being able to utilize these environments fully. GIS will no longer be seen as special stand-alone systems, but will become one of a complementary range of information and analytical facilities that can be made available to users in an organization as required. This will be a key factor in establishing GIS as a mainstream application of IT and achieving many of the goals of the AGI.

The impact of standardization could ultimately extend into the home market through the growth of the mass multi-media market, which will be highly standardized once the market is better established. This will also be affected by developments in telecommunications and the convergence of the entertainment and computing industries which is leading to the creation of 'information super-highways'.

Volume

GIS is currently a low volume, high margin market with a multitude of vendors. The market is, however, likely to become high volume, lower margin with fewer vendors selling more standard products. The impact for the use of geographic information, in general, will be to establish:

- an open market in geographic information

- effective use of geographic information for strategic, as well as tactical, analysis.

It is not suggested that it will be easy to achieve this vision. None of the above will occur without the international market pressure that will emerge from major users of GIS, such as central and local government and utilities. If this does not happen, the GIS market is likely to remain marginalized and will stay outside the wider use of IT.

5.2 Setting priorities for GI/GIS standards

Establishing priorities is one of a number of very difficult GI/GIS standards issues that the AGI needs to address. With unlimited resources it may be possible to make progress on a broad front, but without this luxury opportunities for GI/GIS standards will inevitably have to compete for resources – prioritization is therefore essential.

The prioritization of GI/GIS standards will need to take account of a range of factors, including:

- the logical order in which tasks should be undertaken

- the areas in which standards are realistically achievable

- the views of users and system developers

- the extent of market influence at national, European and international levels

- the expected duration and cost of standards initiatives

- the impact of market forces without standards

- the business benefits of standards

- the cost of implementing standards, by suppliers and users.

5.2.1 **What the UK GIS community thinks**

A brief survey of views from various parts of the GIS community was undertaken to get some indication of what members of the GIS community think about standards generally and what they perceive to be the priorities for standards. Whilst the results clearly were not statistically significant enough to bear close scrutiny, the survey did reveal consistent views on a number of standards issues. A description of each of the areas being considered for standardization is included in Annex A.

Industry issues

In terms of GIS industry issues, the survey suggested that the availability and cost of geographic information is currently the single biggest issue the GIS community faces. GI related standards fell into a second group roughly equal with GIS benefits and business case, and GIS within IS strategy.

GIS system standards, metadata and GIS awareness still registered as significant, but to a lesser extent.

Impact of standards

Just over 50 per cent of respondents felt that standards are generally helpful, with about 30 per cent of the respondents describing standards as essential – a clear signal that GI/GIS standards are widely regarded as important.

Priority areas for standards

The relative importance of GI standards over GIS standards was confirmed by respondents expressing priorities for standards activities. Standard data access and data exchange standards were regarded as of greatest importance, followed by data quality and spatial referencing.

Geographic user interface, portability of applications, GIS corporate integration and data models were regarded as medium priority, followed by concepts and definitions and GIS inter-working.

Levels of standards

National standards were perceived to be most important by all subgroups represented. European standards were felt to be slightly more important than international standards. Note however, that national standards should be specified in procurements of data or systems only where their use cannot be regarded as protectionist.

Funding standards	As far as British standards are concerned, there was a widespread view that government and GIS industry should bear much of the cost of standards making. User subgroups should also play a part.
	Many respondents found it difficult to suggest who should fund European and international standards. The majority of respondents suggested that the European Union should fund European standards.
	The results were inconclusive for international standards.
Role of AGI	Respondents were comfortable with the idea that AGI should promote the use of standards, encourage others to undertake standards development, and promote GIS needs in the development of wider IS standards. However, there was less enthusiasm for AGI being involved in overseeing standards development or raising funds for standards development.

5.2.2 Priority setting

Ideally setting priorities should ultimately be a decision for those professionals who are in possession of all the relevant facts (such as the AGI standards committee). In reality some priorities have to be set by those who are willing and able to sponsor initiatives, often because it is of direct relevance to their field – arguably this is the business driven approach.

5.3 Implementation options

In order to achieve the benefits of GI/GIS standards, there are several quite different approaches that should be considered by those involved in GI/GIS standards-making:

- leave the market to develop its own standards

- support the development of de facto standards

- the de jure approach.

Leave the market to develop its own standards	This is basically allowing the market to develop without any form of direct intervention from the industry. While this is a necessary process at the beginning of a new market, GIS have been available for some years now and are gaining a reasonable share of some markets such as local government.

However, given the barriers to successful use of GIS (see for example the CCTA Information Management Library volume: *Geographic Information Systems in Government: Realizing the Opportunities*) it is unlikely that the market will develop fully without some appropriate standards emerging and there is no evidence that appropriate standards will emerge without some form of guidance from the industry.

Support the development of de facto standards

This approach is modelled on the approach of groups such as X/Open and POSC. It is based on conducting reviews of user requirements and taking existing standards, from industry, de facto or de jure standards, and creating a forum for partnership between users and vendors to implement these standards. This process also involves producing guidelines and publications which help users and vendors to apply standards in an unambiguous way.

De jure approach

This is based on using one of the formal standards setting groups to create the standards, and will be focused on CEN and ISO initiatives. The de jure process tends to be very lengthy and requires significant financial resources, and the ISO and CEN committees would need to anticipate the needs of the developing GIS market.

Key strategy issues

The key issue for assessing the best strategy for the future is not at this stage to attempt to identify specific standards that will become widely used, as the market is not yet sufficiently mature. The aim is to identify the best process which will lead to the development of appropriate standards – whether these are de facto or de jure standards.

There is no reason to believe that the GIS market, if it does develop as anticipated, will differ from other IT-related markets. This suggests that there is scope for all three of the approaches to be employed. GI/GIS standards makers should therefore establish which areas of standards can be left to market forces, which can be effectively resolved by influencing de facto standards, and which areas can only be resolved satisfactorily by formal de jure standards.

The other issue that needs to be considered is whether the priorities for de jure standards development should be data oriented or systems oriented. The authors note that, apart from general IS standards, systems-related work is lagging well behind data-related standards.

Since systems-related standards will almost inevitably be based on de facto standards in the future, it is suggested that systems-related standards should be progressed with groups such as the Open GIS Foundation, X/Open and POSC.

5.4 Conclusion

It is widely understood that members of the UK GIS community wish to pursue a future in which:

- GIS is an accepted part of IS infrastructure, rather than an expensive tool for cartographers, statisticians and other specialists

- useful digital geographic information is widely available, affordable, of good quality and can be easily exchanged between organizations

- GIS is easy to use, inexpensive to buy, low risk and there is a wide choice of systems

- GIS products and applications are compatible with one another, are scalable, and are capable of working within diverse operating environments.

Problems

Each of these developments would be a major step towards the effective use of geographic information. The AGI's standards initiative aims to directly underpin such developments; but whilst the AGI's standards agenda has been largely well defined, it is clear that progress is frustrated by a range of problems:

- lack of awareness of the potential of GI/GIS standards

- limited influence within an international market

- insufficient participation of the right people, at the right time

- severe constraints on available funds

- lengthy timescales for the development of de jure standards

- difficulties in establishing priorities

- costs and delays in the implementation of standards.

It could be argued that these problems suggest that GI/GIS standards are rather easier to talk about than actually to put into practice. If so, what can be done? There are a number of possibilities, that will now be outlined:

- selective standards activity

- monitoring membership needs

- management control of standards initiatives.

Selective standards activity

Whilst it is generally agreed that the scope of AGI's standards agenda has been correctly defined, AGI needs to focus on what is realistically achievable given its limited resources and scope of influence (mainly UK and partly European). In particular it needs to establish which areas can be left to market forces, which can be effectively resolved by influencing de facto standards, and which areas can only be resolved by developing de jure standards.

Given that AGI is unable to realistically fund anything other than very modest standards initiatives, it may be appropriate to concentrate its standards resources and budget on raising awareness and finding sponsors for standards projects.

Bearing in mind that de facto standards generally precede de jure standards, this suggests that the AGI should look particularly at the priorities and requirements for de facto standards. There are already mechanisms in place for establishing such requirements, for example the X/Open Xtra process and surveys carried out by POSC and others. It is, therefore, suggested that the AGI should work with these organizations as far as possible, probably through common members of both X/Open and POSC. This will allow the most appropriate standards to be identified which will gain wide acceptance. These standards may be moved into the de jure process in due course as necessary.

Monitoring membership needs

If AGI is to be selective about its standards activity, it will need to establish priorities. It is suggested that AGI should regularly survey the needs of its members in order to prioritize all its GI/GIS related initiatives, including standards activities. Whilst this could mean sacrificing some standards work in favour of non-standards activities, it would help AGI to target its modest financial resources to areas that are considered most important by its membership.

Where there is evidence of demand for standards activities, as a general principle, those that will directly benefit from the solution should fund it. Having a project 'owner' is a critical success factor for most projects, and the successful standards work within LGMB illustrates the importance of this point.

Management control of standards initiatives

It is also suggested that project management practices should be applied to control standards work. Standards initiatives should have clearly defined terms of reference – a start, a middle and an end. In particular, a business case should be established for individual standards initiatives wherever possible, especially where the funds involved are substantial.

It is not suggested that presenting a credible business case for a standards initiative is ever likely to be easy – estimating costs is difficult enough, and estimating benefits is somewhat harder; but unless those involved in GI/GIS standards are able to justify their work in terms which non-specialists can understand, there is likely to remain little commitment to standards initiatives. The challenge is therefore for prospective standards makers to make their subject accessible and make their case understood; if there is a good case for standards activity or a policy on standards, those who are expected to fund such work can reasonably ask to understand the nature of their investment.

In conclusion, it is hoped that by going some way towards explaining current and prospective GI/GIS standards, this volume has contributed to a greater understanding of the business case for GI/GIS standardization work and of the need for users and vendors to encourage and support this activity. Most importantly, the standards makers are reliant on feedback from users on their needs and priorities.

Annex A: BSI IST/36 – AGI standards agenda

The Association for Geographic Information (AGI) has defined a comprehensive agenda of GI/GIS related standards activities which are described in its *Strategy for Geographic Information Standards*. The strategy includes a range of major initiatives to develop de jure and de facto standards for GI/GIS, and a number of other initiatives to influence general IS standards making and to raise funds. Whilst there has been broad agreement about the scope for GI/GIS standards, there is some doubt and concern about whether the AGI's standards agenda is realistically achievable.

The main areas on the AGI standards agenda are:

- standard data access

- geographic user interface

- data exchange

- portability of applications

- GIS interworking

- GIS corporate integration

- spatial referencing

- data models

- concepts and definitions

- data quality.

Standard data access

To allow conforming GIS products each to access the same databases. This standardization would allow geographic databases that have been created using one GIS to be read and updated by different GIS products. One important result would be to allow a digital map database to be accessed by many different GIS products.

Geographic user interface

To specify how the user interfaces with GIS. GIS which conformed would all have a similar look and feel. This

71

would make it easier for people to switch from one product to another and reduce training.

Data exchange

To allow geographic data to be output from one GIS and input into another GIS product. NTF addresses part of this requirement. Standard data models and dictionaries are required for a complete exchange – see below.

Portability of applications

This would provide a standard programming language/-application environment for GIS. Users could then develop GIS applications suitable for several different GIS products.

GIS interworking

To allow different GIS products to communicate directly. This would allow a general purpose GIS product dynamically to use other specialist GIS products from other suppliers.

GIS corporate integration

To allow two-way communication between GIS and other corporate information systems; for example, alphanumeric databases would be able to access GIS processes and data. This integration would enable the potential of GIS to be made available more easily for corporate management.

Spatial referencing

To ensure that different geographic datasets are spatially consistent and can be directly cross-referenced spatially, without the need for transformation. This would ensure that geographic data sets that had been generated independently could readily be used together for purposes that were not foreseen at the time of creation.

Data models

Standard data models would provide a standard format for particular geographic data sets. The National Street Gazetteer is an example. This process allows the benefit of existing data model design to be passed on and maximizes the potential for transferring data and expertise. Each data model should have a standard data dictionary that describes the format and meaning of each item of data.

Concepts and definitions

A standard set of terms and definitions would provide a common language for those involved with geographic information such as educators, implementors, data or system providers.

Data quality

To enable the ready assessment of the suitability of data sets for purposes other than those for which they were originally produced. This would include the production of appropriate generic assurance and control definitions, guidelines and indices.

Annex B: CEN TC/287 standards agenda

The following text details the main activities proposed to be undertaken by the CEN TC/287 committee, as described in its draft work programme version 6. The work programme is divided into four parts:

B1 Geographic information – overview

This consists of:

- overview

- reference model

- definitions

- dictionary of common terms.

Overview

To provide an overview of the family of standards, to describe the areas of application to which the family applies, and to explain the overall context within which the family will operate.

Reference model

Identification and definition of all important components subject to standardization that are needed to describe, structure, encode, search, order and transport geographic data.

Definitions

To produce a set of definitions for use in standards developed in the field of geographic information.

Dictionary of common terms

To produce a dictionary which supports a common understanding of geographic information and the technology related with it.

B2 Geographic information – data description

The data description aspects cover:

- techniques

- guidelines for application schemas

- geometry

- quality

- metadata

- transfer.

Techniques	To develop or adopt formal languages and graphical notations for defining and describing application data – including geometry-directory and dictionary data. This includes the definition of techniques for mapping from conceptual schemas to transfer schemas.
Guidelines for application schemas	To give guidelines about use of the data description techniques for developing application schemas for geographic information.
Geometry	To define and represent geometric primitives and their constructs in geographic information using the formal languages defined in the work item 'Techniques'. This will include the identification of the types of geometry relationships, the rules for expressing them and the definition of any constraints. It will include non-geometric elements if deemed necessary.
Quality	To define the quality concepts of geographic information, and their integration into dictionary and application schemas.
Metadata	To define the conceptual schema(s) for metadata (directory and dictionary) for geographic datasets. Metadata describes, among others, availability of datasets, information about classification schemas and overall quality.
Transfer	To define transfer schemas and implementation mechanisms for the transfer of geographic data. This will include application data and metadata – both for parts defined by the standard schemas and for parts defined by application specific schemas.

B3 Geographic information – referencing

The referencing aspects cover:

- position

- indirect positioning systems

- time.

	Position	To define how locational reference systems shall be described. Co-ordinate systems should be covered. The choice of any particular positional system is outside the scope of this item.
	Indirect positioning systems	To define how locational reference systems shall be described when the positioning is made indirectly. Non-co-ordinate based systems should be covered.
	Time	To identify how the temporal dimension of geographic data shall be handled in geographic information.
B4	**Geographic information – processing**	The processing section covers query and update. This is to develop or adopt formal languages, including appropriate spatial operators and identification systems, for querying and updating geographic data, including metadata.

Annex C: Comparison of UK with European standards initiatives

(Source: AGI Standards Strategy Report)

CEN TC/287	AGI/BSI IST/36	Comments
Directory and work programme for the European standardization requirements in the field of geographic information.	AGI Standards Strategy Report. GIS Dictionary.	These documents are intended to cover the strategy and framework for standards: AGI projects are either for the national interest or contributions to the European standards process.
Conceptual models for geographic information. Conceptual schema for geographic information. Geographic Information Description Language.	Conceptual, logical and physical data models.	AGI plans work on a conceptual and logical level. CEN will concentrate on more detailed levels.
Quality model.	Introduction to data quality.	AGI is in the process of publishing a guide to data quality.
	Quality assessment of data.	A two-part British Standard is envisaged covering uniform standards for data quality.
	Quality Assurance.	AGI is to make recommendations for the implementation of ISO 9000/BS 5750 schemes.
Positional References. European Referencing System.	Spatial Referencing.	Definition of common direct and indirect spatial/positional referencing systems.
Dataset Definition Language. Geographic Query Language. Update Language. Spatial Operators.	Spatial Data Types. Spatial Operators. SQL Evolution. GIS Query Language.	Some of this work may appear first as de facto standards.
Service.		Anticipates need for EDI messages. Assumes that no work on ISO 8211 or other service standards will be required. (8211 is just completing its revision cycle with BSI and AFNOR participation.)

CEN TC/287	AGI/BSI IST/36	Comments
GIS Development Tools.	GIS Application Programming Interface. Software engineering standards for GIS.	
General Exchange Mechanism. Specialized Exchange Mechanism. Transfer Format. Transfer Profiles.	BS 7567 Electronic Transfer of Geographic Information (NTF). Application Specific uses of NTF. Database Import/Export. European data transfer support project.	National transfer projects exist in parallel with European developments until the latter are sufficiently progressed for reviews to take place. UK has offered to convene a European Working Group on Data Transfer Standards and this has been accepted by TC/287.
Portable GIS.	Framework for Open Systems GIS. GIS Standard Functional Model. GIS Data Manager. GIS Remote Database Access.	
	GIS User Interface. GIS Conformance Testing. GIS Version Control. GIS OSI Application Layer.	(Projects with no European equivalent).
	GIS Benchmark Tests. Available Standards. GIS Standards Roadmap.	(Projects with no European equivalent).
	Development of a business case for Open Systems GIS.	(Projects with no European equivalent).

CEN TC/287	AGI/BSI IST/36	Comments
	GIS Glossary.	(Projects with no European equivalent).
	TIIWG Meta Database.	
	Standard Test Datasets.	
	Spatial Unit Boundary & Attribute Data Meta Databases.	
	National Street Gazetteer.	
	National Land & Property Gazetteer.	
	Postal Address Standard.	

Bibliography

CCTA publications

The following CCTA publications relate closely to, or directly expand upon, topics in this publication:

Information Management Library

The Information Management Library volumes are available from HMSO Publications Centre, PO Box 276, London, SW8 5DT, or telephone 0171 873 9090, fax 0171 873 8200.

- An Introduction to Geographic Information Systems
  ISBN 0 11 330612 1

- Geographic Information Systems: A Buyer's Guide
  ISBN 0 11 330606 7

- Geographic Information Systems in Government: Realizing the Opportunities
  ISBN 0 11 330607 5

- A Survey of Geographic Information Systems in Government
  ISBN 0 11 330635 0

Information Systems Engineering Library

The Information Systems Engineering Library volumes are available from HMSO Publications Centre, PO Box 276, London, SW8 5DT, or telephone 0171 873 9090, fax 0171 873 8200.

- SSADM for handling Geographic Information
  ISBN 0 11 330613 X

IS Planning Subject Guides

The IS Planning Subject Guides are available from HMSO Publications Centre, PO Box 276, London, SW8 5DT, or telephone 0171 873 9090, fax 0171 873 8200.

- Standards and IS Strategy
  ISBN 0 11 330653 9

Other CCTA publications

These publications are available from the CCTA Library, Rosebery Court, St Andrews Business Park, Norwich, NR7 0HS, or telephone 01603 704930.

- Catalogue of Standards for use in IT procurement CCTA/HMSO (no ISBN assigned)

- Guide to the requirements of the IT Standards Decision and the Revised Supplies Directive 1990 CEC DGXIII/CCTA (no ISBN assigned)

- Information Systems Standards – Policy for UK Public Sector Organizations 1993 CCTA/HMSO (no ISBN assigned)

Other publications

The following publications and products relate closely to, or directly expand upon, topics in this publication:

P M Mather (ed)
Progress in Geographic Information Handling – Research and Applications
1993
(Chapter 20: The Impact of GIS on Local Government in Great Britain, H Campbell and I Masser of
Department of Town and Regional Planning, University of Sheffield).
John Wiley & Sons Ltd, Baffins Lane, Chichester, West Sussex, PO19 1UD, or telephone 0800 243407, fax 01243 775878.
ISBN 0 471 94060 7

C F Cargill
Information Technology Standardization: theory, process and organizations
1989
Digital Press, 313 Washington Street, Newton, Massachusetts, MA 02158, USA, or telephone 00 1 617 928 2500.
ISBN 1 55558 022 X

P L Croswell
Open Systems mean 'Open Sesame' for the GIS
community
Article in GIS World, November 1993
GIS World Inc, 155 East Boardwalk Drive, Suite 250, Fort
Collins, Colorado, CO 80525, USA, or telephone
00 1 303 223 4848, fax 00 1 303 223 5700.
ISSN 0897 5507

C D Evans *et al.*
User Needs in Information Technology Standards
1993
Butterworth-Heinemann Publishing Ltd, Linacre House,
Jordan Hill, Oxford, OX2 8DP, or telephone
01865 310366.
ISBN 0 7506 1559 1

P Gray
Open Systems – A Business Strategy for the 1990s
1991
McGraw-Hill International (UK) Ltd, Shoppenhangers
Road, Maidenhead, Berkshire, SL6 2QL, or telephone
01628 23432, fax 01628 770224.
ISBN 0 07 707244 8

D R Green (ed)
Geographic Information 1994
1994
Taylor & Francis Ltd, Rankine Road, Basingstoke,
Hampshire, RG24 8PR, or telephone 01256 840366.
ISBN 0 7484 0071 0

C A Kottman
Some Questions and Answers About Digital Geographic
Information Exchange Standards
1992
Intergraph Corp, USA.
Intergraph (UK) Ltd, Delta Business Park, Great Western
Way, Swindon, Wiltshire, SN5 7XP, or telephone
01793 619999.
(No ISBN assigned)

J Rowley *et al.*
Strategy for Geographic Information Standards
1992
Association for Geographic Information, 12 Great George
Street, London, SW1P 3AD, or telephone 0171 334 3746.
(No ISBN assigned)

GIS Dictionary Version 2
1994
Association for Geographic Information, 12 Great George
Street, London, SW1P 3AD, or telephone 0171 334 3746.
ISBN 0 85406 636 5

Handling Geographic Information. Report on the
Committee of Enquiry chaired by Lord Chorley.
HMSO Publications Centre, PO Box 276, London,
SW8 5DT, or telephone 0171 873 9090, fax 0171 873 8200.
ISBN 0 11 752015 2

Glossary

AFNOR   Association Française de Normalisation – the French national institute for standardization.

AGI   The Association for Geographic Information. The UK umbrella organization for those interested in geographic information and its associated technology.

AGI-BSI IST/36   BSI IST/36 Geographic Information Standards. This is an 'external' technical committee operating under BSI/DISC rules but sponsored by the AGI which pays a composite annual fee to DISC on behalf of its members. Technical committee members need to be either AGI or DISC members. The committee creates UK and influences European and international standards.

ANSI   American National Standards Institute – the US national standards body and member of ISO.

API   Application Programming Interface – the formally defined way in which an application program can interact with an operating system or other system resource.

ASCII   American Standard Code for Information Interchange – a standard binary coding system for representing characters within a computer.

ATKIS   German national format for geographic information exchange.

BSI   British Standards Institution. The UK national standards organization and ISO member.

BS 7567   National Transfer Format (NTF) is the UK standard for the transfer of digital geographic information.

BS 7666   Spatial data sets for geographic referencing in three parts comprising:

        Part 1: Specification for a street gazetteer (1992)
        Part 2: Specification for a land and property gazetteer (1993)
        Part 3: Specification for addresses (1994).

CAD   Computer Aided Design. The design activities, including drafting and illustrating, in which information processing systems are used to carry out functions such as designing or improving a part or a product.

CASE   Computer Aided Software Engineering.

CCITT   Committee Consultitif International et de Télégraphie et de Télécommunications. International organization primarily addressing telecommunications standards.

CEN   Comité Européen de Normalisation. The regional standards group for Europe. CEN is not an internationally recognized standards development organization and so cannot contribute directly to ISO. It functions broadly as a European equivalent of ISO and its key goal is to harmonize standards produced by the standards bodies of its member countries. Membership is open to EU and EFTA countries.

CEN TC/278   This CEN committee is responsible for GDF (Graphics Data File) which has been developed primarily for in-car navigation systems. A draft EN (European standard) is being put forward and is being submitted to the equivalent ISO 204 committee on transport telematics.

CEN TC/287   The main focus of this committee is creating geographic information standards to enable the exchange of data using EDI and/or other means to locate suitable data, identify its quality, construction and cost and then to deliver such data on receipt of an electronic order. This requires standards for metadata, for the reference framework to be self descriptive and encouragement to make data sets widely usable. TC/287 reports to Technical Board BTS 7 which covers Information Technology standards.

CERCO   Comité Européen des Responsables de la Cartographie Officielle. The European Committee of Representatives of Official Cartography, under the auspices of the Council of Europe.

CGM

Computer Graphics Metafile. A standard (ISO 8632) file format for the storage and transfer of picture description (graphical) information.

CGM*PIP

Computer Graphics Metafile*petroleum industry profile.

Client Server

Describes an IT configuration where an application uses the processing power of both a workstation and a host system. The workstation provides an interactive user interface and the host provides large scale data storage and multi-user information sharing facilities.

CORBA

Common Object Request Broker Architecture, a standard developed by the Object Management Group. This provides a standard for objects to communicate across heterogeneous local and wide area networks.

De facto standard

A product or technical feature that has been widely adopted by markets, and is not controlled by a standards body.

De jure standard

A standard that has been officially adopted by a standards organization.

DGIWG

Digital Geographic Information Working Group. A NATO working group responsible for the development of DIGEST.

DIGEST

The Digital Geographic Information Working Group (DGIWG) exchange standard. A NATO standard for the exchange of geographic data in digital form between defence agencies.

DIN

Deutsches Institut für Normung. The German standards organization and ISO member.

DISC

Delivering Information Solutions to Customers. An autonomous part of the British Standards Institution which has responsibility for standards relating to information systems.

Domesday 2000

The Domesday 2000 project is part of the initiative to develop a National Land Information System (NLIS) for Britain by the turn of the century. The project aims to create a national computerized archive of property and land data in Britain. The Domesday Research Group (DRG) was formed in November 1991 under the supervision of Professor Peter Dale to bring together research that could assist the NLIS initiative. The members of the DRG come from both GIS and property backgrounds. The project is likely to be one of the largest proposals for collecting information on land, what lies below it, upon it and above it, not only in Britain but possibly in western Europe.

DXF   Digital Exchange Format. A format for transferring drawings between Computer Aided Design systems, widely used as a de facto standard in the engineering and construction industries.

EDI   Electronic Data Interchange. Interchange of messages/data between computers.

EDIFACT   Electronic Data Interchange For Administration, Commerce and Transport. The ISO 9735 standard stipulates the application level syntax rules. The ISO 8211 standard stipulates the data descriptive file for the messages to be interchanged.

EDIGeO   A data transfer format strongly based on DIGEST, adopted by AFNOR as a French experimental standard.

EN   A European standard.

EPHOS   European Procurement Handbook for Open Systems.

ETAK   A data exchange format used for describing road networks.

ETIS   The European Telecom Information Systems Group.

EU   European Union.

FIPS   (US) Federal Information Processing Standard.

GDF   Graphics Data File – has been developed primarily for in-car navigation systems by major car manufacturers. It is currently being put forward as a draft European standard (EN) and is being submitted to the equivalent ISO 204 committee on Transport Telematics.

GIAG	Geographic Information Advisory Group of the LGMB.
GI	Geographic Information. Information about objects or phenomena which can be related to a location on Earth.
GIS	Geographic Information System. A system for capturing, storing, checking, integrating, manipulating, analysing and displaying data that is spatially referenced to the earth.
GKS	Graphics Kernel Standard ISO 7942 – the original standard for display and output devices, devised for programmers to access standard two-dimensional displays.
GOSIP	Government OSI Profile. A standard defined to simplify the procurement of open systems in the UK public sector. GOSIP describes the standards necessary for open system connectivity.
GPS	Global Positioning System. A constellation of US satellites which enables users with appropriate receivers to fix their position on or above the surface of the Earth to varying degrees of accuracy depending on the receivers and techniques used.
GRASS	A public domain GIS product.
GUI	Graphical User Interface. A user interface that makes use of graphical objects, such as icons, for selection of options, and usually has a windowing capability enabling multiple window displays on the same screen.
HPGL	Hewlett Packard Graphics Language. A commonly used language for controlling pen plotters. It is occasionally used for the storage of graphical data.
IEEE	Institute of Electrical and Electronics Engineers Inc – a major US based international professional body and an accredited US standards setting organization.
IEEE 1003.1	See Posix.

IGES
International Graphics Exchange System. An ANSI standard for the exchange of CAD drawings in digital form.

IGGI
Interdepartmental Group on Geographic Information. A group within central government which considers issues on the use of geographic information. IGGI effectively replaces the former Tradeable Information Initiative Working Group which had a different working agenda.

IHO
International Hydrographic Organization.

IHO S57/DX90
An International Hydrographic Organization working group which published an object catalogue (standard S57) and transfer format (DX90) for vector hydrographic information.

Interoperability
The capability to communicate, execute programs or transfer data among various functional units (items of hardware or software or both, capable of accomplishing a specified purpose) in a manner that requires the user to have little or no knowledge of the unique characteristics of those units.

IRDS
Information Resource Dictionary System. A software tool for controlling, describing, protecting, documenting and facilitating the use of an organization's information resources. Two ISO standards exist: ISO 10027 covers the IRDS framework, and ISO 10728 is the IRDS service interface standard.

ANSI produced different and conflicting US standards for IRDS before withdrawing all resources in 1994 to work on the PCTE standard for repositories instead.

ISDN
Integrated Service Digital Network. An international service which allows a number of forms of communication including voice, video, data and fax.

ISO
International Standards Organization. The international standards body which is incorporated by the United Nations. It is the main de jure international standards setting body.

ISO 7498
See OSI.

ISO 7942	See GKS.
ISO 8211	See EDIFACT.
ISO 8632	See CGM.
ISO 8824	See OSI.
ISO 9735	See EDIFACT.
ISO 9945-1	See Posix.
ISO 9592	See PHIGS.
ISO TC 211	TC 211 is an ISO committee set up in 1994 with a title of Geographic Information (Geomatics) and a scope of geographic information and applications of geographic information.
LAN	Local Area Network.
Layer	In the context of GIS, a subset of digital map data selected on a common theme, but not necessarily containing objects of identical type. For example, a layer may contain road information, including motorways, A-roads, B-roads, service stations, car parks etc.
LGMB	Local Government Management Board.
MAP/TOP	Manufacturing and Automation Protocols – a standard originally specified by General Motors. Technical and Office Protocols – a functional standard for the office environment promoted by Boeing. TOP is built on top of MAP.
MEGRIN	Multi-purpose European Ground-Related Information Network. A CERCO initiative to develop joint topographic data such as national and administrative boundaries.
Metadata	Information about other information. The AGI is responsible for distributing the metadata collected under the Tradeable Information Initiative. This metadata lists major databases held by government.

MIT   Massachusetts Institute of Technology.

Motif   Motif is a graphical user interface developed by OSF for
                        UNIX systems.

Multimedia   The use of text, data, still and motion video, sound and
                        computer graphics by a program to form a composite
                        display.

NATO   North Atlantic Treaty Organization.

NIGIS   Northern Ireland Geographic Information Systems. An
                        initiative to facilitate the exchange of geographic
                        information, between a number of major public-sector
                        organizations and utilities in Northern Ireland.

NIST   National Institute for Standards and Technology. A US
                        government agency that specifies policies on technology
                        purchases including IT.

NJUG   National Joint Utilities Group. An association of public
                        utilities (electricity, gas, telecommunications, water etc.)
                        set up to consider issues of common interest.

NLIS   National Land Information System results from the
                        Citizens Charter initiative and involves HM Land
                        Registry, Inland Revenue Valuation Office, Ordnance
                        Survey and local government. It is similar to Domesday
                        2000.

NTF   National Transfer Format. BS 7567 is a UK standard for
                        the transfer of geographic data, administered by the AGI.

NYNEX   A New York and New England telephone company.

Object   A collection of entities which form a higher level entity
                        within a specific data model.

OLE2   Microsoft's Object Linking and Embedding technology
                        which allows objects to be shared primarily across
                        desktop applications. This will allow interworking
                        primarily within the Windows family of products. The
                        relationship between OLE2 and its component object
                        methodology is still to be fully determined.

Open GRASS Foundation   The organization responsible for placing GRASS in the public domain.

Open Look   A GUI developed by Sun Microsystems and AT&T.

Open Systems   Systems that use standards to enable operation of separately purchased solutions from independent sources of supply for all information handling activities.

Operating Environment   A range of services including operating system, graphical user interfaces, and networks which collectively define the interface between software products and their host computer system.

Operating System   The innate controlling software of a computer.

OS   Ordnance Survey of Great Britain.

OSF   The Open Software Foundation which was formed in 1988 by IBM, DEC, Hewlett-Packard, Apollo, Siemens, Nixdorf and Groupe Bull to develop a version of UNIX as an alternative to AT&T's UNIX System V (SVID).

OSI   Open Systems Interconnection. This defines the accepted international standards (ISO 7498 – 1984) by which open systems should communicate with each other. It takes the form of a seven-layer model of network architecture, with each layer performing a different function. ISO 8824 is an OSI-related standard that specifies an abstract syntax notation. The OSI-related family of standards underpin most information interchange processes and new initiatives.

OSNI   Ordnance Survey of Northern Ireland.

OSTF   Ordnance Survey Transfer Format.

Parser   A software tool that can be used to interpret files of a known syntax.

PDES   Portable Data Exchange Standard. Developed under the STEP programme.

Personal Digital Assistant (PDA)

A hand-held computer that is commonly operated by using a pen or stylus. Some models are able to interpret handwriting.

PEX

PHIGS extension to X. A de facto standard that supports PHIGS on X Windows.

PHIGS

A standard Programmer Hierarchical Interactive Graphics Systems (ISO 9592) set of graphics functions to control the definition, modification, storage and display of hierarchical graphics data.

POSC

Petrotechnical Open Software Corporation. An international non-profit organization with a mission to define, develop and deliver a range of open systems software applications for the petroleum industry.

Posix

'Posix' is the name given to the work and standards originally largely produced by a group of IEEE committees in the USA. These committees are addressing various aspects of creating a standard for a 'portable' operating system including issues such as standard system functions, administration, security and process communication. The US IEEE 1003.1 standard, which specifies the operating system interface, has been adopted as international standard ISO 9945-1. This standard has been widely implemented and is often referred to as the Posix standard. The rest of the 'Posix family' of IEEE 1003.n interrelated standards are in different stages of development and agreement.

SAIF

Spatial Archiving and Interchange Format. A Canadian standard for the transfer of two-, three- and four-dimensional geographic data. Includes facilities for vector and raster data with attributes. It is sponsored by the National Geomatics Standards Committee established by the Canadian General Standards Board.

SDTS

Spatial Data Transfer Standard. The US Information Processing Standard (FIPS) for the transfer of spatial data.

STANAG

A NATO standard.

STEP	A transfer format for graphics data which is currently being developed by ISO to replace·IGES.
SQL	Structured Query Language is an ISO standard (ISO 9075). It is a query language interface for relational databases. SQL is used to define, access and manipulate data stored in these databases. SQL3 is the third major version of SQL and is likely to be ratified in 1995.
TIFF	Tagged Image File Format. A widely used format for the exchange of raster images.
TIGER	Topologically Integrated Geocoding and Referencing. A data format developed by the US Bureau of Census.
Topology	The study of the properties of a geometric figure which are not dependent on position, for example connectivity and relationships between lines, nodes and polygons.
TP	Transaction Processing.
Translator	Similar to a parser, but it generates a file of another syntax. For example, a NTF to Arc Info translator translates from NTF and produces an Arc Info file.
UNIX	An operating system developed by AT&T. Now adopted as an industry standard, allowing the portability of application software between hardware from different manufacturers.
Vienna Agreement	An agreement between ISO and CEN concerning principles of cooperation and collaboration.
WAN	Wide Area Network.
WGS84 co-ordinate system	Co-ordinate system used by Global Positioning Systems (GPS).
X/Open Consortium	An international non-profit making organization which defines, promotes and certifies the conformance of open systems products.
X Windows	X Windows is a device independent graphics system, which allows system developers to build distributed, network transparent, multi-tasking user interfaces. It was

originally developed by MIT in 1984 and has since been widely adopted, particularly by suppliers of UNIX systems. X Windows provides the mechanism for drawing windows, it does not define a particular user interface.

Printed in the United Kingdom for HMSO
Dd 0299892   10/94   C8   3400/3   13110